The "Random Genetic Drift" Fallacy

William B. Provine

To my darling wife,
Gail A. Provine

Contents

Preface   Vii

1. Inbreeding and Outbreeding   1

2. R. A. Fisher and "Random Genetic Drift"   23

3. Sewall Wright on "Random Genetic Drift"   47

4. Experiments on "Random Genetic Drift"   89

5. Kimura and Ohta   131

6. Topics in Population Genetics   161

References   179

Preface

Much of my life, since graduate school, has been devoted to the history of population genetics and its role in evolutionary biology. *The Origins of Theoretical Population Genetics* (1971, 2nd edition, 1991) and *Sewall Wright and Evolutionary Biology* (1986a) were my first books, and I edited with introductory materials *Evolution: Selected Papers* by Sewall Wright (1986b). Renowned population geneticists Ronald A. Fisher and J. B. S. Haldane lived before my time as a serious scholar. Fisher's book, *The Genetical Theory of Natural Selection* (Fisher 1930), I use now, is Wright's copy with his remarks and Fisher's original paper on population genetics (Fisher 1922) was also Wright's copy edited with his detailed comments in pencil. Richard C. Lewontin was my Ph.D. thesis committee member. His study with J. B. S. Haldane became a part of me, too. He always encouraged me to evaluate mechanisms of evolution. This book is the result of decades of my life working on inbreeding, outbreeding, and "random genetic drift" in population genetics, plus many other factors affecting evolution.

Contents of this book

Inbreeding and outbreeding were known to breeders for centuries. Darwin understood both, and analogized evolution in domestic populations to natural populations. The problem for Darwin was heredity – he knew little, and he had no precise understanding of why inbreeding would sometimes destroy a population and other times would not disturb a similar population of the same organism that lived in a different place.

Inbreeding and outbreeding were revolutionized by Mendelian heredity. Walter S. Sutton (1902-3) and Theodor Boveri (1904) tied the chromosomes to Mendelian heredity. The history of inbreeding and outbreeding in Chapter One is one of the best stories in the history of genetics. The rise of population genetics replaced inbreeding by introducing "random genetic drift." Inbreeding lost its hold in small populations and did not have full influence in evolutionary biology. Inbreeding and outbreeding are crucial parts of evolution in the wild and in domestication. One aim of this book is to elevate inbreeding and outbreeding, understood by Mendelian heredity in the years 1904-1924, to new heights of current understanding of evolution of wild and domestic populations.

I evaluate Fisher's views of inbreeding and "random genetic drift" in Chapter Two. Fisher's first paper on population genetics (Fisher 1922) showed his belief that inbreeding (loss of chromosomes) could be modeled by "fortuitous extinction of genes" (I will call this term "random genetic drift" in this book) using a genic F he attached to a chromosome. Fisher's intent was to make "random genetic drift" measure the inbreeding effects in a population, but inbreeding is tied to whole chromosomes and Mendelian heredity, and "random genetic drift" is tied to the locus F. Fisher invented locus F in 1922 to represent any variable in the population including inbreeding and outbreeding, selection, survival of individual genes, neutral genes, Hagedoorns's argument, and many other issues, all in one short paper. Fisher's genic F became a fundamental assumption of population genetics then and to the present.

I met with Wright for a decade. We solved many issues related to his work, both in conversation and by letter, and a few others, we let pass without solution. I was left with no way to solve the confusion of inbreeding (whole chromosomes) and "random genetic drift" (at one locus F) in Sewall Wright's papers, books, or about 120 recorded hours of interviews (housed at the Library of the American Philosophical Society in Philadelphia). I read him notes where he used both inbreeding and "random genetic drift" in the same paragraph. I reminded him that in his famous diagram of his 1932 paper, Seventh International Congress of Genetics, he showed inbreeding happening, yet he wrote paragraphs on "random genetic drift" in his big paper (Wright 1931).

Wright's version of "random genetic drift" was derived by 1925 in manuscript, and in several papers in 1929–1930 before writing his famous (Wright 1931) paper in *Genetics*. Chapter Three traces the rise of Wright as a breeder to one of the best population geneticists in the world. He was the world's best geneticist on inbreeding in 1924, and moved from inbreeding to "random genetic drift" in his theory of evolution. Wright's "random genetic drift" was the same as Fisher's, so their view was called the Fisher/Wright version after about 1950 by population geneticists. Neither Wright nor Fisher endorsed the Wright/Fisher term. J. B. S. Haldane used Fisher's version of "random genetic drift," making it Fisher/Wright/Haldane "random genetic drift."

Wright's interest in Fisher's F was deep. Wright used Fisher's F for all his quantitative work in population genetics, and had exactly the same variables measured by F. Wright's only differences from Fisher were mostly in population size and shape, which do not change the mechanics of Fisher's model of evolution.

Experiments of geneticists on "random genetic drift" on domestic populations between 1940 and 1957 are reviewed in Chapter Four. These experiments took a huge amount of time and were overseen by or with Wright. His confused view of inbreeding and "random genetic drift" give an added look at these experiments and their relationship with inbreeding and "random genetic drift." These experiments were all inbreeding consequences, and none were "random genetic drift." Fisher created a myth in population genetics that F was the key to everything. Meiosis is the key to populations, not Fisher's F.

Chapter Five is devoted to Motoo Kimura and Tomoko Ohta. Kimura began by improving the work of Fisher, Haldane, and Wright in population genetics, and improving the relation of Andrey Kolmogorov, an important mathematic figure in Russia, with previous work of Fisher, Haldane, and Wright. Kimura used Kolmogorov in his equations and made the environment conditions even more tight in a population.

Kimura confused inbreeding and "random genetic drift," just as Fisher, Haldane, and Wright did before him. The same is true for Tomoko Ohta, who became Kimura's colleague in 1967. The neutral theories depended upon "random genetic drift." The role of "random genetic drift" in the neutral theory (Kimura) and the slightly neutral theory or nearly neutral theory (Ohta) occupies the end of this chapter, and tells the story of "random genetic drift" in versions of the neutral theories of evolution.

Chapter Six addresses other issues in population genetics. Topics include: "random genetic drift" in prokaryotes; analysis of "gene pools" used by population genetics; comments on selection of alleles at a locus; statistical physics and population genetics; and final insights on Fisher, Haldane, and Wright.

Chapter One

Inbreeding and Outbreeding

Inbreeding (breeding of related adults), outbreeding (breeding of unrelated adults), and artificial (human) selection have been used by animal and plant breeders going back into unrecorded times. During Charles Darwin's life, animal and plant breeding was booming. Darwin had no actual example of natural selection in nature, so he compared natural selection to artificial selection. He began *On the Origin of Species* by examining pigeon breeding and not until chapter four introduced his version of natural selection. Darwin invented examples to interest the reader. For the rest of the 19th century interest in animal and plant breeding increased. Many institutions hired breeders and invented new plants and animals, and plant breeders such as Luther Burbank in California and Ivan Vladimirovich Michurin in Russia had worldwide reputations. Both were extremely successful in creating new varieties of true-breeding plants before the rediscovery of the Mendel's "laws" of heredity in 1900 by Hugo de Vries and Carl Correns, and soon William Bateson and many others joined the group of those studying inheritance by Mendelian models. Burbank and Michurin continued their work and their fame worldwide in the 20th century without ever engaging or using Mendelian heredity.

Inbreeding and outbreeding in breeding and evolution

Neither Burbank nor Michurin could tell others how much inbreeding was needed to create true breeding species, or outbreeding, or selection. Mendelism was rediscovered in 1900. A young Mendelian, George Harrison Shull, went to observe Burbank so he could understand exactly what Burbank was doing to invent so many new plants. Shull was trained in genetics by his advisor at the University of Chicago, Charles B. Davenport, who moved in 1904 to head the Cold Spring Harbor Laboratory, bringing Shull with him. We shall not examine Shull's (largely failed: see Glass 1980) attempt to understand Burbank's breeding through Mendelism, but examine instead his work on a field of maize (Indian corn). Shull became the first editor of *Genetics* in 1916, moved to Princeton University, and had a distinguished career in genetics. Soon his career would blossom with his expertise in maize.

Within six years of the rediscovery of Mendelism, many experiments on inbreeding and outbreeding had been started, all with the intent of understanding the relationship to Mendel's theory of heredity. After 1902-1904 and the Sutton-Boveri discovery of chromosomes to explain Mendelism, geneticists tried to use Mendelism to conduct experiments on inbreeding and outbreeding. William Earnest Castle, who in 1911 would become the advisor of Sewall Wright at the Bussey Institution of Harvard, started an inbreeding experiment in *Drosophila melanogaster* in 1905, the first experimental genetic use of this organism. Edward Murray East, who also would be Sewall Wright's teacher in graduate school, started inbreeding experiments in maize at the Connecticut Agriculture Station in 1905. The USDA Animal Husbandry Division started in 1906 a major inbreeding and outbreeding experiment with brother-sister mating using guinea pigs. And, the University of Illinois began a selective program for high and low protein content in maize.

East left the Connecticut Agriculture Station to go to the Bussey Institution of Harvard in 1909. He left the inbreeding experiments in maize to his close assistant and soon to be East's graduate student, Herbert J. Hayes, who remained then at Connecticut. They published many papers together on this work. Hayes left the Connecticut Agriculture Station late in 1914 to take a job at the University of Minnesota. East's student, Donald F. Jones, after only one year of graduate school at the Bussey Institution, left to take charge of these inbreeding experiments at the Connecticut Agriculture Station. The work of Castle and East on connections with genetics was the direct background for Wright, who was a graduate student in the Bussey Institution from 1911 to 1915. He left with his Ph.D. to direct the USDA Animal Husbandry Division 1905 inbreeding and outbreeding experiments in guinea pigs from 1915-1925. Wright saw deep implications of these experiments for both animal breeding and evolution in nature.

Shull and inbreeding in maize

Shull reported on his inbreeding and crossbreeding experiments on maize in 1908. Davenport was the "Secretary of the Animal Section" of the American Breeders' Association (ABA). Their yearly reports turned into the *Journal of Heredity* in 1915. Shull, Davenport, Castle, and East came to the 1908 meeting of the American Breeders' Association on January 28-30. When Shull gave his paper, East had an epiphany.

Shull's paper was entitled, "The composition of a field of maize" (Shull, 1908). A field of maize was a highly hybrid population. To find out what was hidden in its variation, inbreeding was required. By selfing his maize, using pollen and eggs from the same plant and breeding them, and using selection for the characters he wished for the maize to exhibit, he soon arrived at "pure lines" of maize. He was using Wilhelm Johannsen's terminology of "pure lines," rather than Mendelian terms. When you breed these pure lines together, you really obtain hybrid maize, and perhaps, he suggested, this was what the farmers should plant. This hybrid maize was not a pure line the farmer could plant again each year, but a very vigorous hybrid maize with sometimes excellent results.

East was at this meeting. He was thinking in terms of Mendelian heredity, not Johannsen's pure lines, and thus of homozygosity in the experimental population. He wrote to Shull:

> Since studying your paper, I agree entirely with your conclusion, and wonder why I have been so stupid as not to see the fact myself. . . . I expect to quote from your paper and add some data of my own in a forthcoming report from this station [Connecticut

Agricultural Station]. (Cited by Donald F. Jones 1944, 223-224)

East and Hayes worked together on hybrid corn and published many papers based on their work at the Connecticut Agricultural Station. They were trying to market the hybrid seeds but East did not inbreed them as much as Shull, and did not get consistent results, nor much interest from farmers. Shull, who believed inbred populations should be nearly homozygous, also discovered his inbred stocks took many years to reach homozygosity. In 1911, Shull tried, but did not publish, starting with four inbred lines, crossing the first two and at the same time the other two. He then crossed the two hybrids that resulted (double-cross breeding). When Jones took over the inbreeding experiments at Connecticut Agricultural Station in 1915 he soon tried to understand the vigor of crossed inbred strains of maize and other organisms. All geneticists understood the experiments by Shull, East, Hayes, and Jones on crossing corn, and by Wright on crossing the inbred lines of guinea pigs. No one, however, had a robust theory of heterosis, the name they used for hybrid vigor.

East, Jones, and Wright

In an extraordinary 1917 paper drawn from his coming Ph.D. thesis, Jones proposed a theory of heterosis (Jones 1917). Jones's theory was that dominant factors were maximized by crossing inbred strains together. Morgan's group at Columbia had been studying the linkage groups with chromosomes in *Drosophila melanogaster* and Jones did this for maize. Jones's Ph.D. thesis presented a theory of production of hybrid seed for maize: the double-cross method of hybrid seed production. Starting with four inbred strains, highly selected, true breeding, and productive, Jones produced fine hybrid seeds in high quantity (East and Jones 1919, 202):

Fig. 42.—Diagram showing a method of double crossing maize to secure maximum yields, illustrated by actual field results.

With all this happening in 1917 and 1918, Jones and East were ready to publish an important book in 1919, *Inbreeding and Outbreeding: Their Genetic and Sociological Significance* (East and Jones 1919). This book is infused with East and Jones's epiphany: Mendelian inheritance has revolutionized our understanding of inbreeding and outbreeding, and how it can be applied to better plant and animal breeding and understanding evolution in nature.

Sewall Wright's graduate education came as East was understanding the problem of inbreeding and production of corn seeds. With both Castle, his advisor, and East teaching there, Wright was much influenced by both. Castle was conducting a selection experiment on hooded rats, and at first, believed that he was changing the Mendelian character of "hooded," a dark head. East, however, doubted Castle's experiment, and Wright dreamed up the crucial experiment. East and Castle both welcomed Wright's crucial experiment which settled the issue in East's favor. Wright was ideal choice for the job at the USDA in charge of the inbreeding experiment on guinea pigs in 1915, in Greenbelt, Maryland (Provine 1986, see chapters 2-5).

Wright knew East and Jones were writing their book on *Inbreeding and Outbreeding* in 1919 and was not surprised to hear from East: would Wright be willing to help them by sending information on the inbreeding and outbreeding experiments on guinea pigs? Wright did a great favor for East by sending him preliminary copies of his crucial reports of 1922, USDA Bulletins 1090 and 1121. The general title of both Bulletins was *The Effects of*

Inbreeding and Crossbreeding on Guinea Pigs; Bulletin 1090 was devoted to decline in vigor and differentiation among inbred families, and Bulletin 1121 to crossbreeding the inbred families. Although East and Jones published their book in 1919, they cited the contents of these bulletins with this comment: "Doctor Wright kindly permitted the authors to read these valuable unpublished papers in manuscript" (East and Jones 1919, p. 277).

Inbreeding and Outbreeding is an exciting genetics book to read (but their racism in the sociological significance is sad). Before coming to examples of inbreeding, they state:

> . . . as we have occasion to emphasize again, the greatest advance in our knowledge of the significance of inbreeding has come through linking its effects with Mendelian phenomena. (East and Jones 1919, 118)

Their maize inbred varieties had been self-fertilized for twelve consecutive generations. East and Jones crossed four inbred varieties, and consistently produced vigorous, large, hybrid maize.

Chapter eight was devoted to "Conceptions as to the Cause of Hybrid Vigor," which East and Jones called "heterosis." Here came a forceful statement of Jones's theory of heterosis. He tied it to the chromosomes of *Drosophila melanogaster* in Morgan's lab, where linkage groups proved to be the chromosomes, with some crossing over in early meiosis but no precise mechanisms. At the end of this chapter, East and Jones again clearly stated what Mendelian heredity had done for understanding inbreeding and outbreeding:

> In tracing the evolution of ideas concerning the effects of inbreeding and outbreeding we must give great credit to Darwin for calling attention to the importance of the phenomena in relation to evolution and for being the first to see that hereditary differences, rather that the mere act of crossing, was the real point involved; but with all due credit to Darwin, it was not until Mendelism became known, appreciated and applied that the first real attack upon the problem was made possible. When linked with Mendelian phenomena it was clearly recognized for the first time that one and the same principle was involved in the effects of inbreeding and the directly opposite effects of outbreeding. Inbreeding was not a process of degeneration. Injurious effects, if present, were due to the segregation of characters. In addition to this segregation of characters the fact was established that an increased growth accompanied the heterozygous condition. All the essential facts were accounted for. (East and Jones 1919, 186)

By the late 1940s, hybrid maize was 78% of planted maize in the USA, increasing to more than 95% by 1960. Improvement of inbred strains made hybrid crosses work well using only two inbred lines.

East and Jones knew immediately that breeding maize was not like breeding guinea pigs, cattle, or horses. Shull and East had settled on hybrid corn as the solution to producing great corn yields.

Inbreeding in corn required self-fertilization; animals generally could not do this, except for brother-sister mating, or parental-offspring mating. Animal breeders wanted true breeds, not hybrids. Animal breeders did not want mammal breeding companies that produced hybrids for farmers.

East and Jones wanted evolution in nature to be illuminated by their understanding of Mendelian inheritance. They had already committed themselves to the extremely important role of artificial selection in production of the inbred lines, eliminating many such lines, and using only the inbred lines for the production of hybrid maize. Understanding this concept, they also knew that crossing of geographically separated populations might be very important for evolution.

"The Rôle of Inbreeding and Outbreeding in Evolution" is the title of Chapter 10. East and Jones pinpointed the difference of sex and no-sex populations: "one reasonable hypothesis to account for everything, *Mendelian segregation and recombination*" (East and Jones 1919, 199). Isolated small sexual, natural, populations have some inbreeding. Mendelism, they argued, paired similar chromosomes in small populations, thus producing inbred unusual variations, most of which were disadvantageous in the natural environment. The result: either extinction, or, less often, a more robust inbred population. Crossbreeding these populations by migration would produce much vigor and variation. Less productive individuals produced fewer offspring, and the more productive more offspring:

... the vital feature in the whole affair, the persistence

in both the animal and plant kingdoms of innumerable mechanisms providing for cross-fertilization, is to be explained solely on the ground of offering selective agencies the greatest amount of raw material. Mendelian inheritance is thus assigned a part in phylogenetic development second only the inherent variability, and the whole history of reproductive change becomes clear without the ill-advised assumption that complex processes like autogamy are harmful in themselves. (East and Jones 1919, 209)

Evolution in nature, with no hint of human contact, proceeds like a livestock breeder or plant breeder. Here again, as in Darwin, we have a view of evolution drawn from plant and animal breeding.

Sewall Wright: inbreeding and outbreeding

In addition to the requirements of USDA to write up the results of the inbreeding experiment in guinea pigs, Wright had other duties as well. He maintained the control population, and crossed inbred populations with each other. He also answered all questions about animal breeding. This constant flow of correspondence took much of Wright's time. His answers were detailed and extremely helpful to those who asked advice (I gave most of Wright's correspondence to the Library of the American Philosophical Society in Philadelphia).

He was also expected to write a USDA Bulletin on the

principles of livestock breeding accessible to farmers all over the USA: a 67 page pamphlet (USDA Bulletin No. 905, 1920) entitled simply: *Principles of Animal Breeding*. Wright then sent a draft of this manuscript to East and Jones. Wright introduced the history of animal breeding, reproduction of animals, hereditary transmission, Mendelian inheritance in many different breeds, heredity of form and function, systems of animal breeding, methods of selection, and the value of purebreds. He told the reader about his work on inbreeding and crossbreeding populations of guinea pigs. Wright let his view of animal breeding show in the very first section:

> It is often believed to-day that successful breeders have some mysterious method of which others are ignorant. Instead, the principles of the successful breeder have been exceedingly simple. He isolates and fixes a good type by careful selection and close breeding [shades of Shull, East, and Jones]. If ambitious to take a greater step in advance, he crosses types with characteristics which seem to offer possibilities for a desirable combination and fixes the new ideal by continued selection and close breeding. He brings inferior stock up to a higher level by consistent use of prepotent sires of the same improved type. (Wright 1920, 2)

The well-informed reader will instantly recognize Wright's "shifting balance theory of evolution in nature" in this passage, the same scheme seen in East and Jones for their chapter on Mendelism and evolution. Populations become small or divided in both views of evolution in nature. Wright's shifting balance theory would occupy a deep place in his view of evolution in nature. He would later argue in detail that his shifting balance theory came from his understanding of animal breeding (Wright 1977); but I would also add plant breeding too, especially in East and Jones, 1919. The essential difference between animal and plant breeding was that self-fertilization worked well in plants and brother-sister or parent-offspring mating was the closest inbreeding possible in animals. The consequence was that maize would be largely homozygous after only 8 to 10 generations of self-fertilization, whereas animals took much longer to reach similar homozygosity.

Working from excellent inbred and genetically relatively pure varieties, Wright's section "Crossbreeding for the Market" suggested that farmers could hybridize these varieties and send the hybrids to market. Farmers did not practice Wright's suggestion because they hesitated to breed cows or hogs for market in the way geneticists produce corn plants.

Another striking factor in this pamphlet is Wright's emphasis and complete understanding of linkage groups, chromosomes, and Mendelian inheritance. In later years, when his work was tied with random genetic drift and a general distribution of genes in a population, Wright paid less attention to linkage groups and chromosomes in quantitative models. In this animal breeding pamphlet, however, Wright's view deserves a full quote:

This phenomenon of linkage has been found to be

very widespread. The first case was found by Professors Bateson and Punnett, of Cambridge University, in the sweet pea. Cases are known in corn and oats, in the primrose and snapdragon, in chickens and pigeons, in mice as well as in rats, in grasshoppers, silkworms, and flies. By far the most thoroughly analyzed case is that of the fruit fly, Drosophila, in which Prof. T. H. Morgan and his coworkers, of Columbia University, have studied hundreds of Mendelian variations. They find that these variations fall into four groups, such that within each group every factor is linked more or less with every other factor, while there is never any linkage between factors in different groups. It is not merely a coincidence that in the fruit fly there are just four pairs of chromosomes.

This statement suggests the accepted explanation of linkage. Factors which are carried by the same chromosome tend to stick together. The chromosomes appear to maintain their identity though all the ordinary cell divisions. Just before the formation of the reproductive cells, the homologous chromosomes come together and twist around each other, giving a chance to an interchange of pieces. The degree of linkage between two factors is believed to measure their distance apart within the chromosome. On this basis Prof. Morgan and his coworkers have actually been able to make maps showing the location of a great number of unit factors in the different chromosomes of the fruit fly, which explains the results of crossing in a very convincing way.

The most remarkable corroboration of the chromosome theory of heredity has been the bringing of the genetic phenomenon of linkage and visible behavior of the chromosome into relation with the solution of the ancient problem of sex determination.
(Wright 1920, 23)

Wright understood well both linkage and sex determination, and its relations with chromosomes and Mendelian heredity, and could explain this to a farmer simply. Part of this understanding is Wright's grasp of some crossing over or recombination in early meiosis. He got this information from the lab of Morgan at Columbia University, but at the same time, he knew that recombination was not understood well at Morgan's lab or by anyone else in the world.

Part of the USDA experiment with guinea pigs began with a population of 23 healthy females and 9 males, who fertilized all the females. Beginning in second generation, all mating was between brother and sister of the same litter. Parental generations were allowed to have offspring through their entire reproductive period, and had to produce a litter with both males and females. If not, this family line terminated. Keepers of the experimental animals chose, under the USDA guidance, the "single best male" and "single best female" in one litter for breeding the next generation, thus making the entire inbreeding experiment a stringent selection experiment also. The guinea pig parents produced between 2.5 and 4 generations per year. Wright's Bulletin 1090 (Wright 1922a) has a picture of four generations of the guinea pigs, but by 1917, 8 of the 23 lines were extinct, and Wright terminated all but 5 the same year. Some of these Wright took with him when he left the USDA in 1925 for the University of Chicago where he bred them until 1955 in physiological biology.

Wright analyzed the many unusual characteristics of the inbred lines, from patterns of coloration, vestigial toes, otocephaly, cyclops, eyeless, and many other deformities. After a long discussion of these characteristics, Wright declared:

> The general conclusion which is suggested by a survey of these abnormalities is the same as that advanced in connection with color, pattern, and vestigial toes. Inbreeding per se has nothing to do with their origin. We find a cyclops, several eyeless pigs, and several head and leg abnormalities among the controls. Whether an abnormality appears in an inbred family depends mainly on its initial heredity in that particular respect. (Wright 1922a, 41)

The three families with no or few deformities had the lowest vigor of any of the lines, and the one with the greatest number of deformities was also the most vigorous during the years 1906-1915. One inbred line was very weak, according to Wright, in the first eight generations, and then became the most prolific. He concluded, "inbreeding seems merely to have brought to light genetic traits in the original stock" (Wright 1922a, 58). Inbreeding had very clear effects in differing the families of the inbred guinea pigs, and Wright understood these effects of inbreeding.

Continued brother-sister inbreeding from chosen individuals in each of the lines insured that chromosomes would be lost every generation and those remaining eventually be paired with themselves or very similar chromosomes. The result was the appearance of deleterious combinations, fixing of some characters and the loss of others, and a general decline in vigor, though in later years, as the chromosomes became more homozygous, the inbreds did almost as well as the control population. In later years, Wright would particularly emphasize this fixing of some characters and loss of others, but he was also keenly aware that intense artificial selection from only the most vigorous could go a long way in keeping the inbred lines from extinction.

When Wright crossed the inbred strains of his guinea pigs, of course, he found an instant solution to problems in the inbred stocks. The hybrids were robust in every way, but further breeding of the F1 generation led to much variation. In both his 1922 papers, Wright brought them as close as possible to the level of inbreeding and outbreeding maize in the work of East and Jones in 1919. Wright's comments on evolution in nature mirrored the comments of East and Jones in their 1919 book. Wright had his idea of his shifting balance theory before them, and shared his papers with East and Jones.

The Hagedoorns, inbreeding, and "random genetic drift"

The mammalian geneticists A. L. and A. C. Hagedoorn, whose favorite experimental organism was the rat, pointed out the effects of inbreeding in natural populations. Castle, his students Sewall Wright and C.C. Little, and geneticist/embryologist Helen Dean King of the Wistar Institute in Philadelphia, thought the genetic work of the Hagedoorns lacked precision (Wright, personal interview). Like Wright, the Hagedoorns had a deep interest in evolution in nature, and used their expertise in mammalian genetics and breeding to infer evolutionary processes. The views of the Hagedoorns on evolution in nature bore a basic similarity to Wright's views in 1931 and 1932, but one would never guess this from Wright's dismissal of their work. The Hagedoorns's book did not cite the work of East, Jones, or Wright.

Fisher read the views of the Hagedoorns in their 1921 book, *The Relative Value of the Processes Causing Evolution*. Their fundamental argument was that Darwinian or artificial selection and Mendelian breeding both caused a reduction of chromosomal variation in isolated populations. An isolated population just had many fewer numbers; this meant they had many fewer of the chromosomes. Selection eliminated inferior hereditary factors stemming from inbreeding, but could not explain most of the observable differences between closely related geographical variations of a natural population in nature. Natural selection caused an inevitable reduction of chromosome variability with only a small proportion of the population produced the next generation, or spawned founder populations in nearby environments.

Chromosomes were eliminated because of inbreeding combined with Mendelian inheritance. In the chapter, "Reduction of variability," the Hagedoorns gave many examples of census size being far larger than actual breeding size, citing evidence from pigs, rats, field-mice, flies, common flowering plants, wheat, and many other organisms. Suppose, they theorized, two very small populations, derived from the same large population, were established on two islands. Soon they would differ consistently from each other.

> The fact that islands are frequently found to have species of plants or animals which exist nowhere else, need not be taken as proof for the adaptation of these species to the conditions on those islands. To explain how all the individuals on one island have come to be pure for one set of characters, we need not ascribe any selection value to those characters. . . .
>
> A group of organisms may become pure for a genotype which causes them to possess some organ or peculiarity, which in their present mode of life is absolutely useless. (A.L. and A.C. Hagedoorn 1921: 123-124)

The Hagedoorns used the equivalent of the production of inbred lines of maize or guinea pigs; inbreeding was less, but had the same basic result.

The Hagedoorns use the term "random sampling" only once in the book: "Even in those cases where colonization is random sampling, the sample will seldom be wholly representative" (Hagedoorns 111). They meant a "random sampling" of whole organisms as a founder population. For the Hagedoorns, inbreeding was the clue leading them to understanding evolution in nature.

What drew Fisher to the Hagedoorns' book was their argument that in small populations of rats on different islands, their favorite example, the differentiation had nothing to do with natural selection. They just differed by their fixed chromosomes as did inbred strains of maize or guinea pigs. Fisher thought that the Hagedoorns underestimated the importance of natural selection in these populations. Fisher disagreed with the Hagedoorns about the role of selection in small inbred populations of rats. Fisher wanted to reject the Hagedoorns' thesis that differences between inbred populations were in any sense random. The Hagedoorns's view of speciation pointed to Moritz Wagner as their intellectual predecessor (Wagner, 1868), but the work of John T. Gulick on Hawaiian snails (Gulick 1888), or many other taxonomists, who clearly believed as the Hagedoorns, suggested most characters of closely related species had little to do with natural selection and were basically random.

Conclusions on inbreeding

Shull discovered that inbreeding maize reveals huge amounts of variation locked in the chromosomes of a population. By the same token, garden peas, as raised by Mendel or breeders, were devoid of variation. They practiced within-pod variation (the pods self-fertilize). East and Jones were simply amazed by the results of application of Mendel to variation, and also devised a theory of evolution based on it. Sewall Wright did that too, probably first.

Wright had no difficulty inventing his theory of evolution in the wild. When he wrote his *Principles of Livestock Breeding* (Wright 1920), he had the same theory of evolution in nature that East and Jones published the previous year, later called Wright's "shifting balance theory of evolution." The USDA experiment between inbreeding guinea pigs and similar experiments on corn gave East, Jones, and Wright the "shifting balance theory" of evolution in nature. Large populations, divided into smaller populations, showed much inbreeding. The similarity of the views on evolution with Charles Darwin's *On the Origin of Species* is striking.

In 1920, most geneticists did not understand evolution in nature, but they did have the clues from inbreeding in domestic populations. Darwin had used this method himself. His understanding had little idea of quantitative methods and understanding of inbreeding in natural settings. In the 20th century, however, Mendelism produced a vibrant inbreeding and outbreeding that clarifies both, and has implications for evolution in domestication and in wild populations.

Wright knew nothing about "random genetic drift" in evolution in early 1924. When Wright wrote his paper in 1920 on animal breeding and presented his section on evolution in nature, he did not mention it. The inbreeding theory was presented by East and Jones in their 1919 book, *Inbreeding and outbreeding: their genetic and sociological significance*. Wright, East, and Jones were on the same track as far as they could see on evolution.

Chapter Two

R. A. Fisher and "Random Genetic Drift"

In 1922 Fisher published an important paper, "The Dominance Ratio," that revolutionized the quantitative modeling of population genetics (Fisher 1922). In the introduction to this 1922 paper, he summarized his earlier paper (Fisher 1918), "The correlation between relatives on the supposition of Mendelian inheritance," as an example of his statistical work in genetics. He likened this earlier work on human heredity to the various causes of the laws of gases:

> . . . the whole investigation may be compared to the analytical treatment of The Theory of Gases, in which it is possible to make the most varied assumptions as to the accidental circumstances and even the essential nature of the individual molecules and yet to develop the general laws as to the behavior of gases, leaving but a few fundamental constants to be determined by experiment. (Fisher 1922, 321-322)

Fisher thought the theory of evolution could be modeled on a Mendelian population with a statistical distribution of genes (which we call alleles) as well as to human heredity. In the laws of gases, gas molecules move randomly. Physicists knew this was untrue, but led nevertheless to very useful laws of statistical physics. Fisher's idea of treating population genetics as a version of physical laws deeply affected the entire history of population genetics (Vladar and Barton 2011). Fisher applied his view on human heredity to evolution in this

1922 paper.

Fisher knew that most genetics was unknown in 1922. He made a sharp divide between *Drosophila* and other organisms:

> In some fortunate circumstances, as in *Drosophila*, it has been possible to isolate and identify the more important of these factors by experimental breeding on the Mendelian method; more frequently, however, and especially in the case of the economically valuable characters of animal and plants, no such analysis has been achieved. In these cases we can confidently fall back on statistical methods, and recognize that if a complete analysis is unattainable it is also unnecessary to practical progress. (Fisher 1922, 322)

Fisher understood the genetics of *Drosophila*, and needed statistical methods to accomplish his aim of creating a quantitative theory of population genetics for other organisms in evolution:

> The present note is designed to discuss the distribution of the frequency ratio of the allelomorphs of dimorphic factors, and the conditions under which the variance of the population may be maintained. A number of points of general interest are shown to flow from statistical premises. (Fisher 1922, 322)

Fisher said "A number of points of general interest are shown to flow from statistical premises," meaning that his statistical physics would determine many points in population genetics. Fisher also introduced both the problem of inbreeding and "fortuitous extinction of genes" ("random genetic drift" in this book) in the introduction to his 1922 paper. He addressed both points in his statistical distribution of genes in the population in evolutionary time.

R. A. Fisher on "random genetic drift"

Fisher started with the Hardy-Weinberg distribution of genes under Mendelian inheritance at equilibrium in one locus, F, which he placed on a chromosome (Fisher 1922). This assumption was his primary contribution to population genetics. He added in selection, the survival of individual genes, assortative mating (which includes inbreeding or outbreeding), and factors not acted upon by selection. Fisher introduced the section dealing with "random genetic drift" as he did other influences on the population: using his F variable.

> The interesting suggestion has recently been put forward that random survival is a more important factor in limiting the variability of species than preferential survival [cites the Hagedoorns 1921]. The ensuing investigation negatives this suggestion."
> (Fisher 1922, p. 323)

The Hagedoorns's studied wild rats that had small population sizes in the winter. Inbreeding occurred as the populations got smaller, the Hagedoorns said, and lost chromosomes. The population size was maybe a hundred individuals or less, and these populations had different qualities each year. The Hagedoorns believed these differences were not adaptive.

Fisher wanted to counter the Hagedoorns's conception of inbreeding in small populations because he believed in adaptation. Fisher invented the thesis, of one Mendelian locus F on a chromosome, that became a tenet of population genetics. F was the center of measures he made in population genetics by using this single locus on a chromosome to model all forces on the population including inbreeding, and all other measures. Every problem had a single locus solution, because it fit his quantitative approach.

Fisher needed a good way to enter inbreeding into his quantitative model. The problem here was his model. He did not know how biology produced "random genetic drift" in the population; he just knew the equation he thought traced inbreeding by his variable F, that he had placed on a chromosome. Fisher was the first person to model "random genetic drift" at a single locus. He recognized that the same process could send some neutral alleles to fixation as well as the rest to extinction.

Everyone in population genetics would follow Fisher. What Fisher had done was to relate inbreeding and "random genetic drift." He used his variable F as the same variable that produced "random genetic drift" and inbreeding. Thus any change in "random genetic drift" would in turn affect inbreeding, or any change in inbreeding also would affect "random genetic drift," and both would operate the same way in populations. Genic "random genetic drift" went the same direction as inbreeding in a small population. When a population became small it would have both inbreeding and "random genetic drift."

Fisher's fundamental problem

Fisher's model assumed that inbreeding of rats of the Hagedoorns turned into "random genetic drift" at a single locus, F, on a Mendelian chromosome. How could one genic locus mimic loss of entire chromosomes in Hagedoorns's rats? The Hagedoorns never used a term other than inbreeding, which had chance results in the rat populations every year. The Hagedoorns would never agree with Fisher's "random genetic drift" at locus F on a chromosome, as being anything like inbreeding in a natural or experimental population of rats or other organisms. The Hagedoorns thought differently than Fisher: if a population was small, it would inbreed, lose chromosomes, and every chromosome would pair with a related chromosome, thus providing loci that were nearly the same, rather than the variation in the original population.

Fisher calculated the rate a neutral gene disappeared from the population by "random genetic drift" was $1/4N$ per generation, where N was the size of the population. The loss of Mendelian alleles by "random genetic drift" was so slow ($2.8N$ generations to halve the alleles in this locus in the population) that Fisher said "random genetic drift" was unimportant in evolution. He did say, in the case of a small colony recently isolated from a very variable species, random survival could be important for a time until extinction, or until the population grew to larger numbers:

> As few specific groups contain less than 10,000 individuals between whom interbreeding takes place,

> the period required for the action of the Hagedoorn
> effect, in the entire absence of mutation is immense.
>
> (Fisher 1922, 423).

Though Fisher invented the idea and quantified the effects of "random genetic drift" on one locus of a chromosome, his exposition was designed to refute the Hagedoorns inbreeding thesis and dismiss the importance of the "random genetic drift" in speciation and in evolution generally. Fisher took an expansive view of N:

> I believe N must usually be the total population of the
> planet, enumerated at sexual maturity, and at the
> minimum of the annual or other periodic fluctuation.
> For birds, twice the number of nests would be good
>
> (Fisher to Wright, August 13, 1929).

What Fisher had done was simply to reduce the inbreeding of the Hagedoorns to a single locus with neutral alleles on one chromosome. He treated every variable force on the population the same way, using F. The Hagedoorns's work on declining populations of rats was inbreeding, involving loss of chromosomes, not the loss of alleles at one neutral locus on one chromosome. Fisher's F is assumed in all population genetics after Fisher 1922.

Fisher, gas, and entropy laws

In 1922, Fisher compared his "law of evolution" with gas laws of physics. Late in this paper, Fisher said, "The distribution of frequency ratio for different factors may be calculated from the condition that this distribution is stable, as is that of velocities in the Theory of Gases" (Fisher 1922, 340), again comparing his analysis of the laws of evolution with the laws of gases.

In his 1930 book, *The Genetical Theory of Natural Selection*, Fisher presented his quantitative theory of natural selection, his fundamental theorem of Natural Selection: *"The role of increase in fitness of any organism at any time is equal to its genetic variance in fitness at that time."* He added:

> It will be noticed that the fundamental theorem proved above bears some remarkable resemblances to the second law of thermodynamics. Both are properties of populations, or aggregates, true irrespective of the nature of units which compose them; both are statistical laws; each requires the constant increase of a measurable quantity, in the one case the entropy of the physical system and the other fitness . . . of a biological population. As in the physical world we can conceive of theoretical systems in which the dissipative forces are wholly absent, and in which the entropy consequently remains constant, though we may not expect to find, biological populations in which the genetic variance is absolutely zero, and in which fitness does not increase. Professor Eddington has remarked "The law that entropy always increases--the second law of thermodynamics--holds, I think, the supreme positive among the laws

of nature." It is not a little instructive that so similar a law should hold the supreme position among the biological sciences. (Fisher 1930, 36-37)

Fisher knew that the biology of populations, however, differs from physical entities and immediately stated the many differences he saw:

While it is possible that both may ultimately be absorbed by some more general principle, for the present we should note that the laws as they stand present profound differences—(1) the systems considered in thermodynamics are permanent; species on the contrary are liable to extinction, although biological improvement must be expected to occur up to the end of their existence. (2) Fitness, although measured by a uniform method, is qualitatively different for every different organism, whereas entropy, like temperature, is taken to have the same meaning for all physical systems. (3) Fitness may be increased or decreased by changes in the environment, without reacting quantitatively upon that environment. (4) Entropy changes are exceptional in the physical world in being irreversible, while irreversible evolutionary changes form no exception among biological phenomena. Finally, (5) entropy changes lead to a progressive disorganization of the physical world, at least from the standpoint of the utilization of energy, while evolutionary changes are generally recognized as producing progressively higher organization in the organic world. (Fisher 1930, 37)

Fisher paid little attention to these contradictions in his thinking about populations and statistical physics, or to the contradictions between his populations and his religious beliefs, as we will see soon.

After discussing *Drosophila* genetics in the Morgan's group at Columbia in 1924, compare Fisher's statements of biological populations, gas laws, and laws of entropy, with his description of chromosomes in a popular paper given in 1924 at the London School of Economics:

> From that time it has been possible to carry out experiments quickly, and statistically sufficient numbers; all the linked factors fall into distinct linkage groups; the linkage relations showed that within each group were arranged in order like beans on a string, so that once their position had been located, the intensity of linkage between any two genes could be predicted; and the linkage groups themselves, or strings of genes, were identified with dark staining bodies, called chromosomes, which can be observed in the nuclei both of the germ cells and of other cells of the body. (Fisher 1924, 190)

Fisher understood chromosomes at this time, yet he modeled inbreeding as "random genetic drift" on one locus on one chromosome. He modeled every variable the same way. He called Mendelism with chromosomes the real solution for understanding inheritance. Fisher knew he had two systems going for the study of organisms. He had Mendelism for chromosomes and statistical physics for his model of populations in nature with no chromosomes. For the rest of his life in evolution, Fisher used both models: statistical physics for his fundamental theorem and Mendelism for his understanding of close linkage on chromosomes.

Fisher's 1934 paper in the first year of the important journal, *Philosophy of Science*, "Indeterminism and Natural Selection" (Fisher, 1934) revealed his views. Population geneticists view Fisher's concept of natural selection as a systematic, deterministic factor in changes of gene frequencies (Vladar and Barton 2011), but this conception is a mistake, and Fisher's belief is easily seen in his 1934 paper. He tried to place his theory of natural selection into the raging debate of determination and indetermination in philosophy and physics. The title is revealing: "Indeterminism and Natural Selection," and the abstract revealed his purpose in writing the article:

> The historical origin and the experimental basis of the concept of physical determinism indicate that this basis was removed with the acceptance of the kinetic theory of matter, while its difficulties are increased by the admission that human nature, in its entirety, is a product of natural causation. An indeterministic view of causation has the advantages (a) of unifying the concept of natural law as in different spheres of human experience and (b) of a greater generality, which precludes the acceptance of

the special case of completely deterministic causation, so long as this is an unproved assumption. It is not inconsistent with the orderliness of the world, or with the fruitful pursuit of natural knowledge. It enriches rather than weakens the concept of causation. It possesses definite advantages with respect to the one-sidedness of human memory, and the phenomena of aiming and striving observable in man and other animals. Among biological theories it appears to be most completely in harmony with the theory of natural selection, which in its statistical nature resembles the second law of thermo-dynamics.

In an indeterministic world natural causation has a creative element, and science is interested in locating the original causes of effects of special interest, and not merely in pushing a chain of causation backwards *ad infinitum*. These contrasting tendencies are illustrated by a critique of the mutation theory, and by an attempt more closely to define the sense in which indeterministic causation should be thought of as creative. (Fisher 1934, *Abstract*, p. 99)

Fisher made a fascinating argument: indeterministic causation would make humans creative and with free will.

Fisher did no research for this paper; he cited only his own papers (mostly his 1930 book) and only one book by Max Planck, *Where is Science Going* (Planck 1932), for which Albert Einstein wrote an introduction. Fisher thought Planck and Einstein believed that beneath statistical physics lay a deterministic basis. The common understanding of natural selection as being determinate was, to Fisher, a profound mistake in biology.

Fisher argued that determinism in evolution renders the concept of causation meaningless: ". . . to the determinist, who perceives the logical consequences of his theory, causation is as non-existent as free-will" (Fisher 1934, 106). Indeterminism, on the other hand, restores human "free-will" and "creativity" in the relation of cause and effect.

> An indeterministic world, then, is one in which the human qualities of aspiration, planning and foresight, are rationally possible. . . . Biologically it might be said that purposive action by the organism as a whole is the crowning stage of an evolutionary process by which relatively large masses of living matter have come to achieve that co-operation of parts and unity of structure which we call individuality. For, on a statistical view of causation, spontaneity or creative causation is at its highest only when perfect unity is achieved. (Fisher 1934, 108)

Fisher believed humans were at the top of evolution. He thought God had put humans into the modern world by means of evolution.

Fisher dismissed "the intervention of personal Creator" because invoking one was "beyond the province of natural science." If humans and other organisms were determined by heredity and environment, and their complex interaction, they could show no free-will, nor purpose or creativity. Fisher's view of causation fit nicely with his biology and his religion, which taught humans received their free will from God.

In a 1950 paper, Fisher evaluated the work of two biologists who wanted more certainty from evolutionary biology, Henri Bergson and Jan Christiaan Smuts, who were trying to find meaning from determination in evolution. Fisher explained carefully that the purpose they wanted was in the indetermination of evolution.

The relation of indeterministic laws in populations of organisms compared to statistical laws of physics had moved to the philosophy of science. Fisher settled the issues in 1934 and apparently found no reason to change his mind, but he was clear about the relation of Christian thought and his views on indeterministic laws in 1950:

> We come here to a close parallelism with Christian discussion on the merits of Faith and Works. Faith, in the form of right intentions and resolution, is assuredly necessary, but there has, I believe, never been lacking through the centuries the parallel, or complementary, conviction that the service of God requires of us also effective action. If men are to see our works, it is of course necessary that they be good, but also and emphatically that they should work, in making the world into a better place. It is not necessary that others should know by what particular

agency the result has been brought about, but to thank God for. We must face the difficulty and responsible task of getting good results actually accomplished. Good intentions and pious observances are no sufficient substitute, and noxious if accepted as substitute. (Fisher 1950, 20).

Fisher did not change his mind about statistical physics, free will, God, or evolutionary biology before he died in 1962.

Fisher was faced by a major problem. He knew inbreeding, but did not have the experiences of the Hagedoorns in actual populations. He was nearly blind and still did much work in biology. Fisher asked himself: what statistical physics would apply to inbreeding in a sexual population? He proposed a neutral locus on a chromosome for inbreeding, and put "random genetic drift" as a genic locus into statistical physics. He did the same thing for all forces on the genome.

Fisher assumed that genic variation at F was the same thing as inbreeding with chromosomes. What follows is a modern look at meiosis, which undermines Fisher's assumption that "random genetic drift" is merely inbreeding.

A modern version of "random genetic drift"

Mitosis is the cell division of normal cells in eukaryotes, and the genome stays constant. Meiosis, in contrast, produces chromosome recombination and two reductions of the chromosomes leading to haploid gametes, that is eggs or sperm with only one set of chromosomes. Eggs and sperm randomly combined is the first stage of Mendelism.

What Fisher lacked in population genetics is now provided in every recent biology or genetics textbook explaining meiosis. The reason Fisher missed this advancement is that meiosis was not widely understood until after 1940. On the back cover of this book is a recent colored diagram of meiosis that gives the modern view of what happens to chromosomes in meiosis.

Meiosis starts with a cell with a full complement set of two chromosomes and produces sperm or eggs, gametes, at the end. The normal chromosome pairs reproduce themselves early in meiosis. Chromosomes line up in a 4-stand phase, and set up crossing over, or recombination, usually at double-strand breaks (DSB), which are generally located between coding DNA. Each chromosome has usually one recombination per arm, or something near 1.5 to 1.8 per chromosome. Sometimes more recombination occurs. After recombination, the 4-strand stage divides to a 2-strand stage, which again divides into gametes with haploid chromosomes. Of the four resulting sperm, each has a full complement of one set of haploid chromosomes; the eggs have their full haploid set also, but only one of four survives in mammals with one large egg like humans. Gametes produced in any sexual organism are limited to the original chromosome pairs of the parent plus any recombination of the chromosomes in meiosis. Gametes do not come from a "gene pool" (I will look at "gene pools" in Chapter Six).

No stage of meiosis produces "random genetic drift." The chromosomes undergo recombination, but no "random genetic drift" whatever. How does "random genetic drift" turn into inbreeding or inbreeding turns into "random genetic drift?" This never happens in meiosis. In inbreeding, the population gets smaller and smaller in chromosomes. How often do the remaining chromosomes recombine? Less often than in large populations because chromosomes are rare, and because these chromosomes are related, so fewer variables are produced by recombination early in meiosis (the only place that recombination occurs). Large populations have lots of chromosomes, and recombination is maximum, but not from "random genetic drift." Inbreeding does not lead to more "random genetic drift" in small populations. No matter what Fisher's locus F does, neither it, nor the chromosomes, produce "random genetic drift" in any size population.

Fertilization between male and female gametes is random, and is normal Mendelian heredity. When a mammalian egg is covered with thousands of sperm, and only one gets through to the egg, the other sperm are out of luck. Biologists call that random breeding. I do not challenge random breeding. Random breeding is not "random genetic drift."

Problems with Fisher's "random genetic drift"

Fisher connected inbreeding to "random genetic drift." He went from the chromosome level (chromosome) right to his F he placed on a chromosome (genic). He also connected whatever happens to his locus F to inbreeding in the population, or vice-versa. "Random genetic drift" was inbreeding to Fisher, who wanted to have a way to measure inbreeding in the population. He could do that with his locus F. We can see from the modern model of meiosis that no "random genetic drift" occurs there. All chromosomes that remain after recombination are complete, then have no more recombination until the next meiotic division. A small population has inbreeding, but no "random genetic drift."

Gametes inherit a haploid set of one of each chromosome from the last stage of meiosis. When a sperm is allowed to fertilize the egg, the new organism has the usual pairs of chromosomes, with one or both chromosomes having recombinations. Meiosis is complicated (you can easily see this in the diagram) but is a great way to recombine and sort in two reductions the chromosomes into gametes.

Recombination sometimes does not happen and then the gametes cannot form because the chromosomes do not separate correctly in meiosis. Geneticists call this nondisjunction of the chromosomes as they pull apart in the two reductions in meiosis making gametes. The recombination of chromosomes must occur to guarantee that the meiosis process works. Recombination still occurs, even with nondisjunction. More can happen to mess up meiosis. Recombination occurs only between non-sister chromosomes; any recombination between sister chromosomes is meaningless because they are the same.

The F that Fisher assigned to a chromosome is absent from this modern view of meiosis. F was Fisher's great addition to population genetics. Fisher used inbreeding as part of a statistical management of the population. No wonder Haldane and Wright used this process to manage a sexual population (and of course many populations give up sex or some portion of it). Fisher also used his F in his equations that he used in populations when he used his fundamental theory. Fisher knew this F was false, but used it to fit his physical statistical equations without chromosomes.

Consider a male and female dog mating. The male makes sperm and the female makes eggs and the two have babies. The babies have no way to inherit the entire genome of either parent, but instead inherit recombined gametes from each parent. The genome of the babies remain the same for the whole life of the babies, with no other process, like "random genetic drift," causing any issue. No recombination of chromosomes takes place except in the early hours of meiosis, which occurs every generation. The genome of any individual in a sexual population must wait for these meiosis chromosomes to recombine and after two divisions form into gametes. Meiosis is the key to understanding sexual populations.

Meiosis is the source of heredity in dogs and any sexual organism. No offspring can change their heredity during their lives, although some mutation can happen from noxious substances. The meiosis diagram on the back of this book shows all the processes that chromosomes go through, but "random genetic drift" does not appear as a meiosis section. Meiosis is the time for any recombination in the

DNA and occurs prior to mating in the next generation. Aside from meiosis, the DNA stays constant.

Eukaryotes, who invented meiosis, experience a fabulous method of producing new recombined gametes. Out of this method geneticists already understand inbreeding in small populations; chromosomes approach homozygosity of chromosomes and show symptoms Sewall Wright found in guinea pigs covered in the last chapter. The evidence points to no "random genetic drift" in population genetics for the same reasons Fisher failed when he tied "random genetic drift" to inbreeding.

Conclusions on Fisher's view of "random genetic drift"

Fisher had no biological explanation of "random genetic drift." After losing other chromosomes from population decline, chromosome pairs produce inbreeding, and all inbreeding was widely appreciated before the invention of "random genetic drift." New recombined chromosomes are drawn from parents, not from "gene pools," as I will show in Chapter Six, where J. F. Crow (1991) showed clearly that both Fisher and Wright assumed the "gene pool." Recombination of chromosomes is maximum in the largest populations, and minimum in smaller populations. Each member keeps its chromosomes until it mates. The chromosomes will be recombined in any later generation, as they were for this generation. Fisher was correct in his analysis of inbreeding, but wrong about "random genetic drift."

Fisher's invention of a single neutral locus F placed on some chromosome for the source of "random genetic drift" was highly praised, and everyone else to the present would follow him. He invented this idea because he wanted to use statistical physics equations in population genetics. He conflated inbreeding and "random genetic drift" in small populations. Fisher's approach, of using one locus of "random genetic drift" on paired chromosomes, did not model inbreeding, yet became fixed in population genetics, so that modern population geneticists include both inbreeding and "random genetic drift" in small populations. Small populations have inbreeding, not "random genetic drift." Large populations have no "random genetic drift."

The modern science of breeding in a sexual population reverses the beliefs of Fisher on "random genetic drift." This giant of population genetics knew little about meiosis and reproduction of natural populations. Fisher never understood meiosis, or that the production of gametes guaranteed a process with no trace of "random genetic drift."

Despite this lack of understanding meiosis, Fisher had established a new population genetics. Fisher did not change his views. Haldane and Wright, along with all population geneticists, accepted Fisher's view. Fisher/Wright "random genetic drift" should be called Fisher/Wright/Haldane "random genetic drift." Fisher, Wright, and Haldane never changed the model Fisher provided in his 1922 paper. Has population genetics changed since 1922? No, population genetics has never changed.

J. B. S. Haldane, dialectical materialism, and "random genetic drift"

Haldane, like Fisher, also wrote an article in the first issue of *Philosophy of Science*. He was an atheist and communist, and had a deep commitment to dialectical materialism. He saw no need for ethics or free will to come from indeterminism as Fisher did above. Haldane encountered quantum mechanics in the late 1920s and thought it was the supreme argument against materialism. A few years later, he was searching for ways to keep materialism in biology, distinguishing it from the uncertainty principle in physics and chemistry. In other words, Haldane was the ideal person to keep dialectical materialism in biology, and I will show this was a tough assignment. Fisher chose the easy way out of this problem, by denying determinism in physics. Haldane countered: "the theory that mind takes advantage of the uncertainty principle is only a form of vitalism" (Haldane 1934, 81).

Haldane gives many reasons why materialism leads to causes in biological organisms. "Evolution appears to have been an escape from the consequences of the uncertainty principle" (Haldane 1934, 83). He had no basis for extending this principle to wider forms of life:

> If bacteria are heated and poisoned with certain reagents, the number of survivors falls off exponentially. This is taken to mean that the life of the cell depends on a single unstable molecule, whose change involves death. As the transformation of such

a molecule involves the uncertainty principle, this principle plays a large part of the life of bacteria. But higher organisms, even protozoa, behave as if their life depended on a number of similar molecules. The uncertainty principle in this form plays a less important part in their lives. They are protected from it by the laws of statistics, just as large material particles consisting of many molecules. (Haldane 1934, 82)

Mind was an even harder issue for Haldane. "Let us compare a human mind with a piece of dead matter. The mind exhibits the phenomenon of purpose." (Haldane 1934, 83). He tried to show how the mind could be produced by the brain and nervous system. He used the uncertainty principle to get from brain to mind, and called it dialectical materialism.

We now think of Haldane and Fisher as being two of the three founders of population genetics. Haldane died in 1964, the very year that the genetic code was fully elucidated. Had he lived until today, he would have known that he had 10 times as many bacteria cells as his own. He would be working on the problem of how organisms have separated from quantum mechanics for over 3.7 billion years, and now he would hold a fully justified materialism in biological organisms. His doubts would be gone.

Haldane on "random genetic drift"

Haldane addressed "finite population, random extinction" in the appendix of his book on evolution (1932).

> The investigation of the case where the total population is finite has been wholly due to Fisher (1930) and Wright (1931). It presents the most serious difficulties you met in this investigation, and indeed some of these have not been solved, but the work has raised some problems of very real mathematical and biological interest.
>
> Let us first consider the suggestion, which has constantly being made, *e.g.* by Hagedoorn (1921) and Elton (1930) that random extinction has been an important cause of evolution. If a population of N individuals possesses variation due to *m* genes, how much of this will have been lost, as we may ask, after *n* generations.

Haldane was tied to Fisher's model. Hagedoorns and Elton were devoted to ecology. They looked at populations that started small, or were remains of a large population that became small from disease or hunting or other causes. Haldane used Fisher's invention: placed one locus with neutral alleles upon one chromosome. Haldane found the same thing as Fisher. He found that random extinction on this one locus loses an allele in 1.39N generations, half of Fisher's calculation in 1922, and Wright encouraged Fisher to change before his 1930 book, which he did. Haldane concluded: "In a numerous species this is a very long period even on an astronomical, let alone a geological time, in other words random extinction plays no part in evolution" (Haldane 1932, 204). Haldane took Elton's objection to his figures, did the job again, and found huge times were still involved. He concluded:

> So random extinction has probably played a very subordinate part of evolution, even in favorable cases.
>
> Other events of the same character, *e.g.* the spread of a new gene from an original single individual to a majority of the species, will require periods of the order of N generations. We cannot say that they have never happened, but we can say that they have played a part quite subordinate compared to that of selection or even mutation. (Haldane 1932, 204)

Fisher and Haldane agreed that "random genetic drift," on a single locus on a chromosome, was a measure of inbreeding in Mendelian populations. Both used the same models and did not understand meiosis, and made a biological mistake. They did not realize their mistake when meiosis, with no "random genetic drift," was understood by 1940. Population geneticists are taught now by other population geneticists, so they still today do not understand the problems with "random genetic drift."

Chapter Three

Sewall Wright on "Random Genetic Drift"

Sewall Wright developed and popularized the concept of "random genetic drift." He did not use the actual term "random genetic drift" until after 1939, instead using such language as "drift at random through the multiple dimension system of gene frequencies" (Wright 1929a, 561), "random shifting of gene frequencies," (Wright 1929b, 287), "random drifting of the gene frequencies," (Wright 1930, 354), or genes "drift at random," (Wright 1931a, 204). I simply use the term "random genetic drift" in this book.

Wright agreed with Fisher in two ways: 1) Using Fisher's model of "random genetic drift," which required F, used by Wright until he died; 2) Inbreeding is equated to "random genetic drift." Fisher used "random genetic drift" to mimic inbreeding, and Wright thought inbreeding caused "random genetic drift," but both agreed that inbreeding and "random genetic drift" were deeply related. Wright contributed a new idea to "random genetic drift:" random sampling of gametes. Wright had used Mendelian heredity many times and now he applied the random sampling of gametes, the first stage of Mendelism, as a producer of "random genetic drift." This chapter argues that none of these reasons produce "random genetic drift."

In my book *Sewall Wright and Evolutionary Biology*, I said clearly: "No copy of the original version of Wright's typescript on evolution in Mendelian populations appears to have survived"

(Provine 1986a, 237). I do now have, however, a copy of his original hand-written manuscript of 68 pages on the back of USDA stationery and will use that version to address Wright's approach in 1925. This document is now at the Library of the American Philosophical Society in Philadelphia.

Wright and Fisher met in Washington in 1924. Wright had not seen Fisher's 1922 paper; they discussed the paper and Fisher said he would send it to Wright immediately. Sewall Wright's edition of that paper, with all his critical comments in 1924, was in my collection. My comments are based on Wright's first impressions of Fisher's 1922 paper. Wright's long paper of 1931b was mostly about his differences from Fisher. They were mostly deleted in the published form because they used correspondence to clear up most of their disagreements before he published his major paper on evolution (Wright 1931b, Provine 1986a). Wright reviewed Fisher (1930) when his book appeared. Here is Fisher's letter to Wright after he read Wright's review:

> I was delighted to see your review of my book in the *The Journal of Heredity*, for August last, which for some reason has only appeared in this country. Your opening paragraphs especially will be most valuable in getting the less genetical sorts of biologists to see that the evolutionary bearings of genetical discussions are not what they were supposed to be; but indeed I ought not to praise one part rather than another for I liked it all heartily. It is in fact the most

understanding review of my book which has yet appeared anywhere, and apart from personal vanity, which will of course absorb any amount of mere praise, that is really what an author craves for. (Provine 1986a, 270. I included all correspondence between Fisher and Wright in my book.)

At this time Fisher and Wright were cooperating in population genetics. That would last at least three more years, when they parted over the evolution of dominance.

Wright's handwritten paper: "The distribution of factor frequencies in Mendelian populations" (1924-25)

Wright read the Fisher 1922 paper and immediately began to write his answer to Fisher, whose name appeared, by my count, 24 times in the 68 pages. Fisher had dismissed any evolutionary impact of "random genetic drift" based on his neutral locus F on a chromosome in the population. Fisher invented "random genetic drift" to model inbreeding. Wright could have objected to Fisher's model of one locus with neutral alleles to represent inbreeding. Wright nevertheless accepted Fisher's model for "random genetic drift" on one locus of a chromosome because Wright also wanted to have the population genetics of Fisher. For the rest of Wright's life he had both inbreeding and "random genetic drift" in small populations, as did Fisher and Haldane.

Wright accepted also Fisher's attempt to allow all factors that determine gene frequency in a population to be Fisher's F and its relation to all the other alleles. Wright accepted everything about Fisher's model, but not the population size Fisher advocated. Wright then started a campaign to argue for large populations divided into small population sizes in opposition to Fisher's emphasis upon large populations without sub-populations. Wright's handwritten "paper" was his first chance to lay out his opposition to Fisher, but at the same time he was totally in Fisher's camp as a quantitative population genetic expert. He and Fisher worked out all their problems of population genetics models (Haldane too) and by 1932, their opposition had become arguing about population size, not their population genetics models to which all agreed. Wright wrestled with the problem of having both inbreeding and "random genetic drift" in small populations. He never solved this problem, nor did Fisher or Haldane. He used his inbreeding coefficient (F) as the same coefficient for "random genetic drift," as seen in his major volumes on evolution (Wright, 1968, 1969, 1977, 1978a).

Wright's handwritten paper in 1925 was his first answer to Fisher, and the first paragraph of his introduction said:

> It has been obvious since the Mendelian mechanism became known that the relative frequency of allelomorphs in any population should tend to remain constant in the absence of selection or mutation. It is not to be expected, however, that there will be absolute constancy in any population of limited

size. Merely by chance one or the other of the allelomorph may be expected to increase its frequency in a generation and in this time the population may drift a long way from the original values. The decrease in heterozygosis following inbreeding is a well known statistical consequence of such chance variation.

(Wright manuscript)

Wright never escaped the biological causes of "random genetic drift:" inbreeding, random sampling of gametes, and following Fisher's mathematical model of "random genetic drift."

Wright argued for a large population, divided into small populations, and gave many examples in the rest of the manuscript. He also challenged several of Fisher's assertions in his 1922 paper, the most important being Fisher's rate of loss of genes by "random genetic drift:" $1/4N$. Wright figured the reduction was $1/2N$ and he wondered where Fisher had made a mistake. Fisher admitted the mistake in correspondence and Wright admitted in return that he also made a mistake on selection, making their mathematical models with Fisher's F the same. They did not have any major issues on their models after 1932.

Wright on inbreeding and the effects of random sampling of gametes upon "random genetic drift"

After 1925, Wright had definite ideas about the biological causes of "random genetic drift." Fisher already made the mistake of turning inbreeding into a statistical physics model of "random genetic drift" at one locus with neutral alleles. Wright accepted his model, both mathematical and the interaction of inbreeding and "random genetic drift." His two reasons for "random genetic drift" follow. The first begins:

> Dr. Fisher is interested in the figure 1/2N, measuring decrease in variance, only because of its extreme smallness, from which he argues that the effects of random sampling are negligible in evolution (except as bearing on the chances of loss of a recent originated gene). I, on the contrary, have attributed to the inbreeding effect, measured by this coefficient, an essential rôle in the theory of evolution, arguing that the effective breeding population . . . may after all be relatively small compared with the actual size of the size of the population. (Wright 1930, 352)

Wright missed here, perhaps, that Fisher had used the same argument, the direct connection of "random genetic drift" to inbreeding. Fisher concluded that "random genetic drift" was a measure of inbreeding. Wright has just concluded that the effects of inbreeding produced "random genetic drift."

The second reason for "random genetic drift" from Wright's review of Fisher's book follows:

> I would not deny the possibility of very slow evolutionary advance through this mechanism but it has seemed to me that there is another mechanism

which would be much more effective in preventing the system of gene frequencies from settling into a state of equilibrium, than the occurrence of new immediately favorable mutations. If the population is not too large, the effects of random sampling of gametes in each generation brings about a random drifting of the gene frequencies about their mean positions of equilibrium. In such a population, we can not speak of a single equilibrium value but of probability arrays for each gene, even under constant external conditions. If the population is too small, this random drifting about leads inevitably to fixation of one or the other of the allelomorph, loss of variance, and degeneration. At a certain intermediate size of population, however (relative to prevailing mutation and selection rates), there will be a continuous kaleidoscopic shifting of the prevailing gene combinations, not adaptive itself, but providing an opportunity for the occasional appearance of new adaptive combinations of types which would never be reached by a direct selection process. (Wright 1930, 85)

Wright's second view on "random genetic drift" is described in the middle of this paragraph. "The effects of random sampling of gametes in each generation brings about a random drifting of the gene frequencies about their mean positions of equilibrium" producing a "continuous kaleidoscopic shifting of the prevailing gene combinations." Wright continued this confusion from his two views on the origin of "random genetic drift" for the remainder of his life, and all the time he believed in Fisher's quantitative F on the chromosome. Wright believed that random sampling of gametes in Mendelism produced "random genetic drift" at every locus in small populations, and also that inbreeding led to rampant "random genetic drift" in small populations, in the same paper or book.

Wright wanted a more expansive view of "random genetic drift" making it more important in evolution than Fisher imagined. Wright simply extended "random genetic drift" at one locus on the chromosome as in the Fisher model to every single locus on every chromosome in a large population with sub-populations. Philip W. Hedrick, a famous population geneticist, has described Wright's view on "random genetic drift" perfectly in his book, *Genetics of Populations*, and he accepted Wright's argument:

> All of the above examples of restricted population size can have the same general genetic consequence: a small population size causes chance alterations in allelic frequencies. The random change of allelic frequencies that results from the sampling of gametes from generation to generation is called **genetic drift**. Genetic drift has the same expected effect on all loci in the genome. (Hedrick 2000, 229)

Wright used this qualitative view of "random genetic drift" in his own work for the rest of his life. Wright generalized "random genetic drift" from its Mendelian chromosome model of a single locus to "random genetic drift" at every locus on every chromosome in populations of the right small size.

Wright followed Fisher's model on "random genetic drift." Both his causes of "random genetic drift" are a combination of inbreeding and random sampling of gametes, neither of which causes "random genetic drift." Wright never changed his mind about his causes of "random genetic drift;" neither did Fisher nor Haldane change their views. Fisher had won the real war between them, and his views now persist in population genetics.

Wright's problems with "random genetic drift"

Wright's more formal development of "random genetic drift" came in his big and influential paper, "Evolution in Mendelian Populations" (Wright 1931). He was self-consciously following Fisher's 1922 paper by constructing his own statistical distribution of gene frequencies deduced from his method of inbreeding. Using one locus, two allele models only, he discussed factors that cause gene frequencies in F to vary. He first addressed mutation pressure, migration pressure, selection pressure, and equilibrium under selection (such as with heterozygote superiority) following Fisher. Wright then wrote a section on multiple alleles at a locus:

> The foregoing discussion has dealt formally only with pairs of allelomorphs, a wholly inadequate basis for consideration of the evolutionary process

> unless extension can be made to multiple allelomorphs. Among the laboratory rodents some 40 percent of the known series of factors affecting coat color are already known to be multiple. The number of multiple series is large in other organisms, for example, Drosophila (Morgan, Sturtevant, and Bridges, 1925). It is not unlikely that further study will indicate that all series are potentially multiple. (Wright 1931a, 106-107)

Wright made sure the reader understood that each locus probably had very many alleles.

Wright, as Fisher and Haldane did, made chromosomes disappear in his quantitative version of evolution. When the chromosomes disappeared, inbreeding, based upon chromosomes, took a strange presence. Wright knew the effects of inbreeding, caused by interaction of related chromosomes. By 1931, Wright often took the view that inbreeding produces "random genetic drift" at every locus on every chromosome, even though inbreeding related only to whole chromosomes, which disappeared every generation, and disappeared in Wright's equations, just as Fisher's interpretation of F.

In this big paper on evolution, Wright has now declared that inbreeding leads to "random genetic drift," which affects every locus on every chromosome in the population of organisms. Thus Wright has so far made three arguments:

1) Wright used Fisher's mathematical model, of genic F placed on a chromosome, for "random genetic drift" and all causes of change at one locus.

2) Inbreeding leads to "random genetic drift" at every locus on every chromosome in small populations or large populations with small subsections. Wright used this argument for the rest of his life. Fisher, who invented this model, used it too.

3) Random sampling of gametes, the first stage of Mendelian inheritance, leads to "random genetic drift" at every locus on every chromosome, with maximum drift at the smallest size of population. In his shifting balance theory of evolution in the wild, Wright argued that large populations were divided into small populations, so a constant shifting of each locus in the population was affected by "random genetic drift."

Each of these three arguments need further discussion, and here is the first one. Fisher's model of a locus F placed on a chromosome promoted "random genetic drift" in small populations. Wright used this model also, but included large populations divided into small populations. Wright knew that his populations of guinea pigs were so special that a loose guinea pig could be placed back into the proper population, but he then concluded that inbreeding had caused the variation in each population. Wright later would believe that "random genetic drift" caused this variation. The fundamental conclusion is that many small populations show much variations between populations, produced by inbreeding, but no "random genetic drift." No argument supports any "random genetic drift" in populations of any size in Fisher and Wright, when we think of meiosis instead of the supposed variations in F. Inbreeding explained the genetic variation in small populations because the chromosomes were related, and produced inbreeding results.

Inbreeding, Wright's second reason for "random genetic drift," put related chromosomes together, yielding the populations that Wright viewed in his guinea pigs. Fisher's view also took this position. How could inbreeding cause "random genetic drift?" Inbreeding is determined by loss of whole chromosomes, whereas "random genetic drift" is defined by loss of genes at a locus.

Random sampling of gametes was Wright's third argument for "random genetic drift." In meiosis, recombination of chromosomes produce different gametes, which happen every generation. But meiosis produced no "random genetic drift" in the population for direct reasons in the previous chapter on Fisher. Wright's use of random sampling of gametes, the first part of Mendelian heredity, does not lead to "random genetic drift" in the populations in Fisher, Wright, or Haldane. Wright did not see "random genetic drift" except for smaller populations. Random sampling of gametes, however, produced no "random genetic drift" of any size of population.

Wright's concept of "random genetic drift" remained deeply related to his understanding of inbreeding effects. Wright did understand later that Fisher used "random genetic drift" as a quantitative measure of inbreeding. "Random genetic drift" at each locus increased as population sizes decreased, and genetic variation began to disappear from the population, just as loss of chromosomes stemmed from inbreeding. The loss of heterozygosis from "random genetic drift" paralleled the loss of chromosomes under inbreeding in small populations from Fisher's model. Wright's section on "random genetic drift" in his big 1931 paper was immediately followed by a section titled, "Rate of decrease in heterozygosis," in which he used his inbreeding coefficient to measure its loss of chromosomes. Inbreeding, of course, was related to the loss of chromosome variation. Wright would later explicitly extend his inbreeding coefficient in 1939 to be the rate of "random genetic drift." His confusion of "random genetic drift" with inbreeding effects was then complete in the 1950s continuing to his death in 1988.

Between 1932 and 1937, Wright himself was the primary promoter of his view of "random genetic drift." In 1932, at the 7th International Congress of Genetics held at Cornell University, Wright, Fisher, and Haldane appeared in the same session. At the USDA, he really worked at communication with farmers. He used that ability in 1932, and wrote perhaps his most accessible paper on evolution. In this paper, Wright revealed the effects of "random genetic drift" (labeled as "inbreeding") in natural populations (Wright 1932, 361):

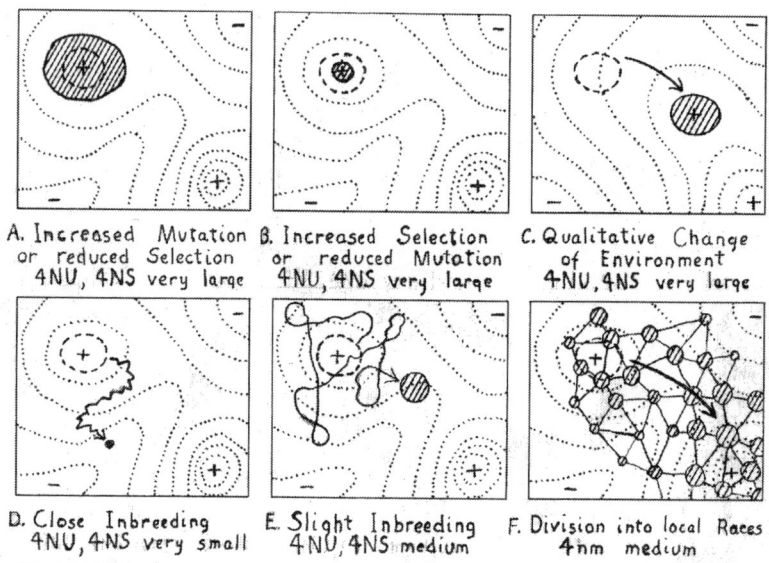

Wright's 1932 paper given at Cornell. I copied his original.

Wright's famous diagram, shown at the 1932 conference at Cornell, gave his ideas a graphic boost. The panels D, E, and F all concern inbreeding. Close inbreeding led to extinction (D). Slight inbreeding was extremely slow (E). Division of a large population into small populations was the ideal for maximizing evolution in nature (F), because inbreeding resulted in novel variants no longer good for the peak on the fitness surface, but by chance adapted to a neighboring peak. This deme could selectively diffuse the entire population by sending out migrants to the other small populations as in (F). Wright's "shifting balance theory of evolution in nature," still the object of much controversy, dominated his theory.

Wright's shifting balance theory required the kaleidoscopic shifting of all gene frequencies at every locus on every chromosome in small populations. How could this occur in a population with little genetic variation? The shifting balance theory executes best with only inbreeding effects, rather than with "random genetic drift" at every locus on every chromosome. East and the Hagedoorns would have no problem using inbreeding for their views of evolution. Wright's labeling of his famous diagram in 1932 was wholly "inbreeding" rather than "random genetic drift," and his title lacks the title of any version of "random genetic drift." Inbreeding has real effects, with which Wright totally agreed.

"Inbreeding" and "random genetic drift" were related in Wright's view from 1932 until his death at 98 in 1988. Wright uses "close inbreeding" in D above to signify that the population "random genetic drifted" itself into extinction. Wright, in his very revealing paper "The Relations of Livestock Breeding to Theories of Evolution" (Wright 1978b), states that inbreeding and "random genetic drift" have the same effect:

> My studies of closely inbred lines of guinea pigs revealed the profound differentiations brought about in all respects by the cumulative accidents of sampling [of gametes] expected at all heterallelic loci under such inbreeding. (Wright 1978b, 1196)

Wright revealed his actual position. Inbreeding affected every chromosome in the population, and affected sampling of every locus in the population. Wright was using the same coefficient for inbreeding and "random genetic drift." I have argued before that this view of inbreeding and "random genetic drift" is not viable in small populations for Fisher or Wright.

Wright wrote his paper on the USDA guinea pig inbreeding experiment in 1922, and he invoked inbreeding to explain the results. He saw no need for this extra mechanism of "random genetic drift" in 1922. In 1978b, inbreeding had been replaced by "the profound differentiations brought about in all respects by the cumulative accidents of sampling [of gametes] expected at all heterallelic loci under such inbreeding." Wright thought all hetero-allelic [modern spelling] loci on every single chromosome were affected by "random genetic drift" and inbreeding at the same time. Wright wanted his method to justify Fisher's model. Wright found two mechanisms that seemed to fit this problem: the random sampling of gametes, and inbreeding, and then used Fisher's model. I argue in this chapter that neither of Wright's two mechanisms, one being Fisher's inbreeding and the other Wright's random sampling of gametes can produce any "random genetic drift."

Observed small populations show nothing but inbreeding to me. I could not understand why inbreeding was somehow replaced with "random genetic drift" in small populations. This problem was solved by reading Fisher, Wright, and Haldane on this problem. They did not show how "random genetic drift" in a small or large population could be possible.

Wright's random drift applied to evolution

Wright stated clearly in his 1932 paper that the nonadaptive differences observed by the majority of systematists between closely related subspecies, species, genera, and even sub-families could be explained by nonadaptive "random genetic drift."

> That evolution involves nonadaptive differentiation to a large extent at the subspecies and even the species level is indicated by the kinds of differences by which such groups are actually distinguished by systematists. It is only at the subfamily and family levels that clear-cut adaptive differences become true (Robson, Jacot). The principal evolutionary mechanism in the origin of species must thus be an essentially nonadaptive one.
> (Wright 1932, 363-364)

The Wright 1932 paper represents his strongest view of the evolutionary mechanism as non-adaptive.

At the level of local races, Wright at last provided the key to the observations and judgement of systematists:

> Subdivision into numerous local races whose differences are largely nonadaptive has been recorded in other organisms wherever a sufficiently detailed study has been made. Among the land snails of the Hawaiian Islands, Gulick (sixty years ago) found that each mountain valley, often each grove of trees, had

> its own characteristic type, differing from others in "non-utilitarian" respects. Gulick attributed this differentiation to inbreeding. More recently, Crampton has found a similar situation in the land snails of Tahiti and has followed over a period of years evolutionary changes which seem to be of the type here discussed. I may also refer to the studies of fishes by David Starr Jordan, garter snakes by Ruthven, bird lice by Kellogg, deer mice by Sumner, and gall wasps by Kinsey as others which indicate the role of local isolation as a differentiating factor. Many other cases are discussed by Osborn and especially by Rensch in recent summaries. Many of these authors insist on the nonadaptive characters of most of the differences between local races. Others attribute all differences to the environment, but this seems to be more an expression of faith than a view based on tangible evidence. (Wright 1932, 364 in Provine, ed. 1986b)

Inbreeding and the "random genetic drift" caused by it, according to Wright in this paper, were the key to understanding the non-adaptive differences between closely allied local races. Wright changed evolutionary biology much more than he expected. After his papers of 1931 and especially 1932, taxonomists, evolutionists, and geneticists could not think anymore about a small population without "random genetic drift." Wright had used the Fisher model, with his addition of random mating of gametes and inbreeding producing "random genetic drift," but this simulation in model was still the same as Fisher's model, as Fisher noted many times.

A founder population is a small migration to a new isolated area. According to Wright, founder effects carry this name because of "random genetic drift" toward extinction or possibly to a more wide-spread species. Instead, a founder population is merely like any small population that experiences inbreeding. Depending upon levels of homozygosity, more or less inbreeding effects will happen to the founder population, which will go extinct or become widespread. Most small founder populations go extinct, same as for small inbreeding eukaryote populations everywhere.

Wright's version of "random genetic drift," so strong in founder populations, inbred populations, and in small populations, or large populations with division into smaller populations, comes to dominate evolutionary biology. Wright, Fisher, and Haldane were now allied in theory. All three accepted Fisher's version of "random genetic drift" as tied closely to small populations, although Fisher and Haldane thought these populations were bigger than did Wright.

Wright and "random genetic drift" in speciation

In 1940, Wright published two important papers on speciation. One he presented at a joint symposium on speciation, chaired and organized by Ernst Mayr, at the annual meetings of two organizations: The American Society of Zoologists and The American Association for the Advancement of Science, and later published in the prestigious *The American Naturalist* (Wright 1940a). The second paper on speciation appeared in Julian Huxley (editor), *The New Systematics* (Wright 1940b), the most prestigious location available in England.

Wright wrote both papers on speciation soon after his major monograph, *Statistical Genetics in Relation to Evolution* appeared in France (Wright 1939). This 62 page monograph was Wright's major publication on all components of his shifting balance theory and speciation. This monograph was never widely read. The French publisher issued it just before WWII broke out in Europe, and the monograph was never widely distributed after the war. Both of the 1940s papers Wright wrote with this monograph clearly in mind. Wright saw speciation as a direct result of his shifting balance theory:

> Fine subdivision of a species into partially isolated local populations provides a most effective mechanism for trial and error in the field of gene combinations and thus for evolutionary advance by intergroup selection.
>
> Complete isolation of a portion of such a species should result relatively rapidly in differentiation of specific rank, to a large extent non-adaptive but adaptive in so far as there has been differential selection or as primarily non-adaptive changes turn out to be preadaptive. Such isolation is usually geographic in character at the outset, but may be clinched after long continued separation by a

gradual chance accumulation of genic and chromosomal differences that in combination bring cross sterility. (Wright 1939, 324)

Wright believed that "random genetic drift," in the sense of providing a kaleidoscopic shifting of gene frequencies at every locus in every chromosome, is the key to speciation, but again he was following Fisher on his model. For Wright, speciation is largely non-adaptive at this time of life. Wright cites many taxonomists who think that differences between closely allied species are non-adaptive. The first paragraph of Wright's summary says it all:

The central problem treated is that of the probability distribution of gene frequencies, tending on the one hand to approach an equilibrium under the pressure of opposing forces and on the other to drift at random under accidents of sampling. It is brought out that accidents of sampling may be much more important in populations in which interbreeding is restricted than is at first apparent. (Wright 1939, 336)

Wright often used "accidents of sampling gametes" as his primary argument for "random genetic drift" in small populations, meaning that variation produced is great in large populations and small in small ones.

Making inbreeding and "random genetic drift" the same force on small populations was a biological mistake. In 1922, Wright, East, and Jones all shared a robust theory of speciation. They were arguing that inbreeding had just the result of speciation when a large population was subdivided into small populations where inbreeding was intense.

Wright's 1940a paper, "Breeding Structure of Populations in Relation to Speciation," which he delivered in Mayr's symposium on speciation in 1939, is crucial for his later work on speciation. Mayr never said a bad word about Wright's work until 1959, well after Mayr published his paper about founder effects and speciation (Mayr 1954). Mayr liked this paper that he quoted from Wright's paper in his paper for the same symposium, and said that Wright had exhibited "the fewest flaws" of any species definition. (To see how Wright and Mayr parted company, see Provine 1986a, 479-484.)

Wright's 1940a paper had a major influence upon many biologists interested in speciation. His theory of speciation affected how both Mayr and Dobzhansky thought of speciation; determined the views of Warren P. Spencer on speciation; and impressed George Simpson and Ledyard Stebbins. Wright had seen inbreeding in the guinea pigs with brother/sister mating, but he never saw any speciation of the inbred lines, even when he crossed the lines. Wright was clear in his mind: populations were small within a larger one, then either grew or went extinct. His drawing expressed this idea in a famous diagram below. Successive generations of individuals go from left to right, and new types appear as shown, with mutations indicated by crossing vertical lines:

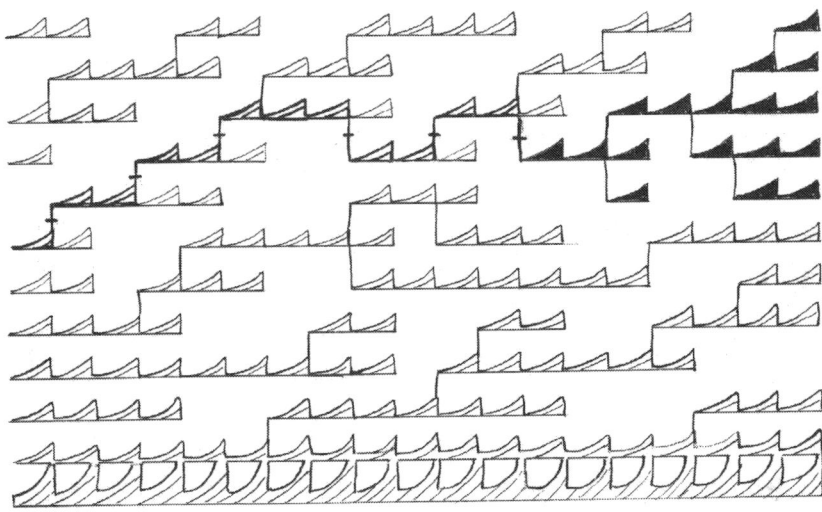

Diagram is from Wright's original version, used by Wright 1940a.

This diagram was the origin of Mayr's "founder" populations in his 1942 book, though Wright later would complain that Mayr took only a single founder effect, not the many more demonstrated in his diagram. Successive founder events gave, according to Wright, a better chance of speciation than a single population bottleneck.

Summing up, we have attempted to show that the breeding structure of populations has a number of important consequences with respect to speciation. Partial isolation of local populations, even if merely by distance is important, not only a possible precursor of splitting of the species, but also as leading to more rapid evolutionary change of the population as a single system and thus more rapid differentiation from other populations from which it is completely isolated. Local differentiation within a species, based either on the nonadaptive inbreeding effect or on local conditions of selection or both, permits trial and error both within series of multiple alleles and between gene combinations and thus a more effective process of selection than possible in a purely panmictic population. (Wright 1940a, 359-360).

Wright's emphasis on a "nonadaptive inbreeding effect" might lull you into thinking that it led to speciation; but no, it "permits trial and error within series of multiple alleles" or "random genetic drift," which can lead to speciation perhaps in combination with natural selection.

Wright's theory of speciation was not helped by his addition of "random genetic drift" to inbreeding effect. Wright could talk about nonadaptive speciation, with inbreeding playing a prominent role, as Wright saw in his USDA guinea pigs. The addition of "kaleidoscopic shifting of gene frequencies at every locus on every chromosome" caused by "random sampling of gametes" is a mistake. Wright's small populations had no "random genetic drift" and he had weakened his theory of speciation.

Dobzhansky and dissemination of Wright's "random genetic drift"

Present at the session with Wright, Fisher, and Haldane at Cornell in 1932 was Theodosius Dobzhansky, who later said he "fell in love" with Wright immediately. Wright's presentation and published paper deeply influenced Dobzhansky. Known for his prolific output and intensive research, he wrote the initial and perhaps most widely read book of the "evolutionary synthesis" period, the three editions of *Genetics and the Origin of Species* (Dobzhansky 1937, 1942, 1951). Dobzhansky gave Wright's work a prominent place in the book, relying heavily upon both "random genetic drift" and Wright's shifting balance system of evolution in natural populations, and applied both to speciation.

In chapter 5, "Variation in Natural Populations," Dobzhansky began by presenting Mendelian heredity and the Hardy's (Hardy-Weinberg) equilibrium that preserved variability in ideal populations. Natural populations were not ideal–and here he introduced "random genetic drift." More accurate historically than most evolutionary biologists today, who cite only Wright on early "random genetic drift," Dobzhansky cited the Hagedoorns, Fisher, Wright, Dubinin, and Romashov as his progenitors on "random genetic drift." His model was drawn from his former compatriots in the Soviet Union, Dubinin and Romashov. They invented the use of dishes to represent the "genofund" (gene pool; see Adams 1979) for a particular locus, and filled them with marbles or beans. Use of a dish (now usually a jar) to represent the "gene pool" of a locus meant Wright's model of "random genetic drift" was being followed. The "gene pool" (covered in Chapter Six) became part of population genetics soon after Dobzhansky introduced this idea in early 1951.

In *Genetics and the Origin of Species* (1937), "random genetic drift" took the form of the usual elementary presentation in a modern biology or evolutionary biology textbook. Western biologists read rather little of Dubinin and Romashov, so their work, as well as Wright's, was filtered through Dobzhansky, whose presentation was clear. He also emphasized strongly Wright's concept of effective population size N, which was generally much smaller than observed population size (shades of the Hagedoorns). If N were large, Hardy's equilibrium dominated. If N were small, then "random genetic drift" caused extinction of some alleles, and fixation of others, until the population almost became homozygous. Dobzhansky accepted Wright's version of "random genetic drift" (kaleidoscopic shifting of allele frequencies at every single locus on every chromosome), and at the same time he used the Fisher model when trying to be quantitative. Neither the first (1937), second (1941), or third (1951) editions of Dobzhansky's book even mentions linkage in a chromosome in the index, though he was certainly familiar with linkage and studied chromosomes daily. In his extension of this book, *The Genetics of the Evolutionary Process* 1970, linkage appears only once in the index. Dobzhansky thought linkage was insignificant in evolution when he used "random genetic drift."

What were the evolutionary consequences of "random genetic drift?" Dobzhansky explained clearly the connection between "random genetic drift" and non-adaptive differences between closely related species or races:

> The conclusion arrived at is an important one: the differentiation of a species into local or other races may take place without the action of natural selection. A subdivision of the species into isolated populations, plus time to allow a sufficient number of generations to elapse (the number of generations being a function of population size), is all that is necessary for race formation. This statement is not to be construed to imply a denial of the importance of selection. It means only that racial differentiation need not necessarily or in every case be due to the effects of selection. (Dobzhansky 1937, 134).

Dobzhansky gave many illustrations of "random genetic drift" in his discussions of variation in natural populations (his chapter V) and used his exciting new work on *Drosophila pseudoobscura* as one of his examples of "random genetic drift" and selection:

> The present writer is impressed by the fact that his [Wright's] scheme is best able to explain the old and familiar observation that races and species frequently differ in characteristics to which it very hard to ascribe any adaptive value. Since in a semi-isolated colony of a species the fixation or loss of genes is to a certain degree independent of their adaptive values (owing to the restriction of the population size), a colony may become different from others simultaneously in several characters. One or a few of the latter may be

adaptive, and may enable the population to conquer new territories or ecological situations. The rest of the characters may be neutral with respect to adaptation, and yet they may be spread concomitantly with the adaptive ones. For example, the chromosome structures that are so variable in *Drosophila pseudoobscura* can hardly be regarded as anything other than neutral characters, although some of them have become racial characteristics in subdivisions of the species populations. (Dobzhansky 1937, 190-191)

Throughout the book, following Wright, Dobzhansky emphasized that evolution depended upon a balance of factors and assessed "random genetic drift" as a central factor in this balance.

Dobzhansky, even before publication of the book in 1937, had embarked upon an ambitious series of papers on the "Genetics of Natural Populations" (GNP). In wild populations of *Drosophila pseudoobscura*, Dobzhansky, using an organism whose ecology was a total mystery to him and others, decided from observation that population sizes were very small, so differences in frequencies of chromosomal inversions between populations must be caused by "random genetic drift." The combination of his book and these prominent papers on genetics of natural populations gave much attention to "random genetic drift" and effective population sizes.

Dobzhansky wanted his experimental work to yield data that could enable someone more sophisticated than he to deduce N (along with other population parameters), thus yielding a robust guess about the importance of "random genetic drift" in local populations. He induced Wright to collaborate in what later became five papers in the GNP series. Starting with the first, GNP #5, Dobzhansky's high hopes for determining N through his data were much diminished. Wright's analysis, very tentative by his own account, gave no precise estimate of N, but Nm together (where m is the immigration rate). Since no independent estimate of N or m was possible, population sizes could only be estimated. Wright suggested that N was somewhere between 500 and 2500, but these figures were far larger than Dobzhansky imagined, thus reducing the influence of "random genetic drift" as an important factor in measured differences between local populations. Wright was moderating the constant use of "random genetic drift" as the major explanatory factor, according to Dobzhansky, in the earlier four papers in the GNP series.

Under Wright's influence, Dobzhansky did indeed moderate his enthusiastic initial take on the importance of "random genetic drift" in evolution in nature, but never relinquished his emphasis upon the possible effects of "random genetic drift" in his writing about evolution. Dobzhansky did a fantastic job of spreading Wright's view of "random genetic drift" by emphasizing that small populations will be drifting at every locus on every chromosome, an idea that went back to Fisher. Students of evolution and genetics using Dobzhansky's *Genetics and the Origin of Species* would get a strong dose of Wright's version of "random genetic drift."

Julian Huxley, H.J. Muller, and Sewall Wright: zoo populations and "random genetic drift"

The question of "random genetic drift" had much practical interest then as now regarding small populations in zoos. Muller, a Nobel prize winner, who was at the University of Edinburgh in 1938 after leaving the Spanish Civil War and before that, the USSR; Huxley, polymath of evolution and Director of the London Zoo; and Wright exchanged letters about the dangers associated with the small size of zoo populations, and the size required for a species to survive for many decades instead of a few years or generations. The application of Wright's ideas on "random genetic drift" governs the whole discussion:

Julian Huxley to Sewall Wright, 10th October 1938

> Dear Dr. Sewall Wright,
>
> Muller was staying with me on his way through to Edinburgh last week, and said that he had a recollection of your having stated that there was a definite minimum size for an isolated population below which it could not indefinitely maintain itself, on account of the random accumulation of deleterious mutations. I asked him to give me the reference, as I had no recollection of it, and he writes that he thinks it must have been in some talk of yours.

If you could let me know about this, I should be interested. It follows, I think, from your general statements, but I wanted to know whether you could give any actual figures—though of course these would differ from case to case.

The matter has a practical as well as a theoretical bearing in regard to conservation. E.g. in this country there are only about 7 pairs of Kites left, and a great deal of money is being spent on watching their nests and conservation measures in general. If the idea of a minimum is correct, it follows that this alone can be of no avail in the long run, and that we should import foreign stock to cross with them. This is at present being opposed by the ornithological purists on the ground that it would mix up subspecies: However, if the alternative is extinction, they may agree.

A similar case arises with regard to the northern group of White Rhinoceros in Africa, etc.

There are many cases which present great theoretical interest, where well differentiated but extremely small populations exist, e.g. on islands. For example, the census of the St. Kilda subspecies of Wren, which is very well characterized, showed that the total population was about 70 pairs. In such cases I take it that there probably is occasional chance immigration of Wrens from the mainland which are still capable of interbreeding and introduce "new blood". If the rate of immigration was of the same order as the rate of deleterious mutation, the harmful effect would be, I presume, avoided.

I should much like your views on these points.

H.J. Muller to Sewall Wright, 20th October 1938

Dear Wright,

Huxley has sent on to me a copy of the letter of October 10th which he sent to you. I should perhaps explain a little further, saying that what he has referred to is a *numerical* value for the population size below which degeneration will occur through the process of drift. Of course I know well enough that you have calculated the value in relation to the values for mutation rate, migration, and selection, but this result is an algebraic one. It seems to me that somewhere, some time, you made certain "reasonable assumptions" regarding the numerical values which the latter quantities might have and in that way arrived at a number of about 1,000 for what might be called the minimum population size capable of indefinite perpetuation. I think that the numbers assumed had been so extreme that the result could be regarded as a minimum in the sense that it was very unlikely that the real number would be lower.

However I have not been able to find such a statement by you in print anywhere, and I am wondering whether I imagined it.

I did not at the time know the exact problem that Huxley had in mind. Of course, the effect is such a long-time one that only negligible damage would, I should think, be done by allowing the reduced numbers mentioned to continue during the period when present political and economic conditions prevail. Is it not true that, if 7 pairs of Kites were later allowed to multiply rapidly and form a large population, in a few generations they would probably recover completely from the effect of the reduction? On the other hand, once mixed with a different variety, they could never be unscrambled, although I cannot quite see the object in trying to preserve all accidentally formed geographical races of other species now existing. Past ages were not so kind to us, if that may be called kindness.

Sewall Wright to Julian Huxley, October 31, 1938

Dear Dr. Huxley:

Theoretically there should be a certain minimum size below which a population cannot maintain itself in the long run (*Genetics* 16, p. 139-

142, 1931). In a large random breeding population an equilibrium should be established between the pressures of recurrent deleterious mutation and selection, such that the most favorable allele in each series is always the most abundant.

In such a population, gene frequencies will vary only slightly from the equilibrium point. With decrease in the size of population, random deviations become larger and one allele or the other drifts into fixation. Since the type genes are more numerous they are much more likely to be fixed than their deleterious alleles but fixation of only a few of the latter may be disastrous to the line. This is an immediate effect of small numbers. But since the effect of selection is largely eliminated below a certain relation to size of population ($4NS$ less than 1) the mean gene frequency shifts gradually from the above equilibrium point to a point determined solely by opposing mutation pressures. This inevitably means a further lowering of adaptive values. The second process is, of course, very slow since it depends on the accumulation of deleterious mutations.

Assume 100 recessive genes each of which gives a 10% loss of reproductive efficiency ($S = .10$). Assume that each arises by mutation at the rate $u = 10^{-5}$ per generation and that reverse mutation occurs at the rate $v = 10^{-7}$. In a large population the equilibrium point is at $1 - \sqrt{v/s}$ or 99% type to 1% recessive genes. The frequency of recessive *individuals* is only .01% for each and thus only about 1 individual in a 100 has any one of 100 such deleterious genes. In a small population there will be random fixation, the chance of fixation of the type gene being .99 and of the mutant .01. With 100 such loci, each individual will have on the average 1 deleterious recessive instead of .01 as before. This may bring the birth rate below the death rate and so lead to rapid extinction without any new mutations.

If not, such mutations will accumulate. If extinction does not intervene, the average frequency of type will approach $\sqrt{(v/u+v)}$ or about .01 (instead of .99). The average individual will have some 99 deleterious factors to 1 type (instead of 1 or .01 of the preceding cases).

All of this refers to the fate of a *single* population that remains small and subject to the *same* conditions of selection for a long period. An additional danger comes from the reduced variability of a small population and hence low elasticity in meeting changing conditions. In my experiments with inbreeding in guinea pigs (Tech. Bull. 103, U.S. Dept. Agr 1929), many strains ran out rapidly presumably as a result of the immediate fixation of injurious factors. Even the 5 most vigorous strains as a group fell well below the control stock in all elements of vigor during the first nine years. Analysis of a second 9 year period indicated no further decline, on the average, (indicating that fixation had occurred but that the time was too short for any appreciable accumulation of mutations). At any given time there were clearly significant differences among the inbred strains in fecundity and mortality percentages. These differences did not however remain the same throughout the 9 years (in contrast with such characters as weight). For two or three years a certain strain would seem to be capable of multiplying indefinitely while in another strain it was necessary to make every possible mating to avert extinction. A few years later the situation might be reversed. Such a change occurring simultaneously in all branches could not have been due to mutation to an important extent. The most probable interpretation seemed to be that the changes reflected changes in conditions. One strain was better adapted to one set of conditions, another to another. The random bred stock could produce individuals adapted in any of these conditions. Thus no matter how well adapted an isogenic strain may be to particular conditions it is likely sooner or later to encounter conditions which it cannot meet and become extinct.

It is hardly possible to put any actual figure on the minimum number for a given species. It would seem very unlikely that a kite population could be maintained for many years a level of less that a dozen pairs. A catastrophe is almost certain to occur in some year apart from any biological deterioration. If, however, the number could be increased rapidly, the population might not suffer from having passed through such a small number. In Dr. Allee's recent book (*The Social Life of Animals*) he compares two cases of species that passed through periods of small numbers. The Laughing Gull was reduced to 12 pairs restricted to Muskegat Island in 1880. Careful protection led to increase in numbers until now there are many thousands and it has spread in large numbers to the mainland. The Heath Hen also was reduced to very small numbers and restricted to one island (Martha's Vineyard) at about the same time. There was an increase in numbers under stringent protection until some 2000 were estimated in 1916. A second reduction in numbers in 1917 due to a succession of unfavorable conditions was never recovered from, in spite of strenuous efforts and no bird has been seen since 1932.

We can see from this fascinating correspondence that by 1937, Huxley, Muller, and Wright no longer use "inbreeding" and talk only about "random genetic drift," as the important factor in small populations. Wright does, however, go right back to the inbred guinea pig experiment to explain to Huxley and Muller what might happen in zoo populations, but by this time, Wright no longer used "inbreeding" to explain the effects of inbreeding, but instead, used "random genetic drift." From this time onward, invocation of "random genetic drift" in small populations became the standard of population genetics and evolutionary biology, and biologists used "random genetic drift" for talking about speciation.

The genetic code was not deciphered until two years after Fisher's death, and the same year Haldane died. Fisher and Haldane missed out on the revolution that affected study of evolution: DNA, RNA, proteins, and their complex biochemical interaction which depended upon biochemical machines. Wright lived on to age 98 (he died in 1988), but his opinions and the equations he used changed little after the mid-1950s, despite his four volume work on evolution (Wright 1968, 1969, 1977, 1978a).

Wright, laws of physics, and "random genetic drift"

We have seen in the previous chapter that Fisher and Haldane both published in the *Philosophy of Science* in 1934. Wright's first paper in a philosophy journal came in 1964, the year of Haldane's death and two years after Fisher's death. A fascinating work could be written about how their papers appeared at all in a philosophy journal, and Wright's story would involve Charles Hartshorne, his dear friend in Philosophy at the University of Chicago. As Wright says in the *Introduction* to his paper, "Biology and the Philosophy of Life":

> I should state at once that I am writing as a biologist, specifically as geneticist, interested in philosophical implications of his subject, but with only a superficial knowledge of philosophy in general. (Wright 1964)

I refused to see any connection of Wright's views on philosophy of biology to his work in population genetics when I wrote *Sewall Wright and Evolutionary Biology*. Now I must revise. Wright was caught in the same problem as Fisher and Haldane on human free will in relation to biology and physical equations. Fisher chose religion, Haldane chose materialism, with holes, and Wright chose dual-aspect or monistic panpsychism (Wright 1964, 280). Wright saw biology and physics and chemistry as one, and that any material body had a corresponding psychism that was hidden from biologists. To Wright, the entire natural history of neural states in evolution and corresponding psychism was never covered by biology.

Wright developed this view in 1914 when he read Karl Pearson's *Grammar of Science* and completely misunderstood Pearson's argument for using science to study mind. Pearson described the work of William K. Clifford, who advocated a monistic panpsychism, that Pearson dismissed. For more than 30 years, Wright cited Pearson's book as the source of his views on panpsychism. Finally, he did some research for his 1964 paper and rediscovered that the author he relied upon was Clifford rather than Pearson, and the whole picture came clear for him:

> It must be admitted that in discussions through the years I have found few biologists who seem at all attracted to this view. It was accordingly very gratifying to find far reaching agreement with Charles Hartshorne who had arrived at a similar view by a wholly different route. In Whitehead's *Science and the Modern World* (1925) to which I was now led, I found the most complete philosophical development of this point of view although I cannot say that I can penetrate into the full depth of his meaning.

Fisher was in classes with Whitehead as teacher at Cambridge University. Wright and Hartshorne spoke frequently at the University of Chicago and they often talked about Whitehead and Clifford. This paper by Wright is more interesting because he, like Fisher, wanted for humans to have free will and could not find it in deterministic or indeterministic biology.

In his section titled, *Freedom of the Will*, Wright began:

> Subjectively, mind seems to involve the continual exercise of choice, of course, a limited range of possibilities. How is this freedom to be reconciled with apparent determinism in the behavior of non-living things? Part of the answer is that is as already noticed that the laws of nature are merely statistical and no more preclude choice on the part of the

> individual components than do statistical laws of human behavior. . . . Freedom of the will is sometimes considered to be equivalent to caprice, but mere chance is as little compatible with freedom of will as determinism. (Wright 1964, 285).

Fisher found the key to free will in indeterminism and god, and Wright rejected that, too. The mind behind the view of biology, monistic panpsychism, somehow made the free will possible. Fisher, Haldane, and Wright were confused about the laws of physics and biology. The problems were really tough. They did not know enough about biology or physics to make these difficulties go away.

Conclusions on Wright and "random genetic drift"

Would Wright's shifting balance theory have suffered much if he had stuck with inbreeding effects instead of his "random genetic drift"? In my humble opinion, the shifting balance theory would have been much improved by subtracting Wright's "random genetic drift." With just inbreeding in small populations, his shifting balance theory would gain power, lose nothing, and oppose Fisher. But Wright would have reasons to reject population genetics by using Fisher's model. He could have expressed his shifting balance theory in his 1920 paper on animal breeding. East and Jones did so in 1919, using only inbreeding effects in small populations. Wright had only to add "large populations subdivided into much smaller but still connected populations" as drawn in panel F of his diagram in the 1932 paper.

For Wright, his commitment to the tie of "random genetic drift" with the shifting balance theory lent weight to his theory, and he certainly never changed his mind.

Chapter Four

Experiments on "Random Genetic Drift"

Beginning in the 1940s, a new emphasis on quantitative experiments on "random genetic drift" in laboratory-controlled populations grew. By 1957 the results of this work settled the issue for geneticists and evolutionists with regard to "random genetic drift" in small populations over tens or hundreds of generations. Except for Wright, participants in this work did not talk about inbreeding, but only about "random genetic drift." In the years of inbreeding experiments, 1906-1922, geneticists found inbreeding in small populations but no "random genetic drift." Starting with new experiments in 1940, inbreeding had disappeared and small populations had only "random genetic drift."

"Random genetic drift" for Fisher was inbreeding. Wright had moved quickly to make inbreeding and "random genetic drift" the exact same thing. His papers with Kerr in this chapter shows how much he equated them. We already know from Chapter One that inbreeding was known to all geneticists. Wright said with Kerr they were the same thing, but inbreeding is at the chromosome level, and "random genetic drift" is at the genic level. While inbreeding, the population would have fewer chromosomes; Wright equated few chromosomes with increased "random genetic drift."

Wright's previous experiments with guinea pigs worked well as an experimental population, and he understood inbreeding. To keep up with Fisher, Wright needed to have "random genetic drift" that works in inbreeding populations, or small populations in large populations. Wright concluded by equating inbreeding and "random genetic drift." In a small population, this is biological nonsense. Meiosis produces no "random genetic drift" (see chapters Two and Three).

Although no evidence existed in 1940, population genetics experts expected to have excellent experimental evidence for "random genetic drift" in this work on experimental populations. The production of work, mostly by graduate students in the 17 years examined, is presented in this chapter.

In 1951, Arthur J. Cain wrote to Wright with a particularly insightful comment. "I have been trying to work out a suitable experiment to *demonstrate* drift, but it seems very difficult." (Cain to Wright, April 4, 1951). Cain was here referring only to sampling drift of gametes. The 1950s would become the era of the demonstration of sampling "random genetic drift" by laboratory experiments using *Drosophila*. The analysis presented in this decade was decisive. Beyond a general invocation, which happens everywhere in population genetics and biology, few other experiments demonstrating "random genetic drift" in eukaryotes have occurred in the laboratory since 1957.

"Random genetic drift" has replaced inbreeding in evolutionary biology and population genetics. Experiments were tried by wonderful students, who could not tell inbreeding from "random genetic drift," and their advisors, who taught their students Wright's "random genetic drift." Each of these experiments had a true explanation: inbreeding. Wright thought that inbreeding led to great strength of "random genetic drift" in small populations. Wright was biologically correct about inbreeding and wrong about "random genetic drift."

Warren P. Spencer on Drosophila and random drift

Although given many opportunities to leave College of Wooster, Spencer elected to spend his career there. His teaching duties were demanding. He also maintained his research on wild Drosophila at night and in the summers. Spencer looked for evidence of "random genetic drift" and Wright's shifting balance theory in his wild Drosophila (Baker, W. K, Gregg, T. G., Neel, J. V. and Stalker, H. D. 1975).

In the late 1920s, Spencer began research on wild populations of Drosophila that he brought to his lab. No geneticist understood much about Drosophila ecology at this time, with Alfred Henry Sturtevant a possible exception. Spencer wanted to connect his understanding of Drosophila ecology with mutations he observed. He could not find much correlation but thought that mutations were probably produced by environmental factors rather than internal factors (Spencer 1935).

Spencer had read Wright and heard him speak. His research on wild Drosophila changed to connect ecology of his flies with population genetics. In 1940, Spencer published the paper "On the Biology of *Drosophila immigrans Sturtevant* with Special Reference to the Genetic Structure of Populations" (Spencer 1940). The "genetic structure of populations" is code for "Sewall Wright's views." His conclusion in the paper was that population size was the crucial factor determining genetic structure, citing Wright (1931). He suspected huge changes in population size over a single year; "random genetic drift" occurred when the population size was small.

Spencer's paper on "random genetic drift" appeared in the first issue of *Evolution* (Spencer 1947), entitled "Genetic Drift in a Population of *Drosophila immigrans*." To the work reported in Spencer 1940, Spencer added two more years of data from 1944–1946; he recovered many recessive mutations by inbreeding the flies. He found that some mutations gained frequency in these two sample years in comparison with the 1940 report.

Spencer's samples of equal size taken at the same season of the year and in the same way, but separated in space by almost a quarter mile and in time by two years, showed the genes stubble, brick, and dubonnet had gained a relatively high frequency in the *Drosophila immigrans* population of this Western Pennsylvania village.

> When we consider the breeding structure of *Drosophila immigrans* populations in this latitude, the most reasonable explanation of the high frequency of these genes is that of random genetic drift, fluctuations due to chance rather than any concentration of genes due to high mutation frequency or selective advantage. (Spencer 1947, 108).

What exactly was the breeding structure? Huge fluctuations of population size occurred every year. Summer populations were large, but winter populations were very small indeed. Spring founder populations, very small, would naturally experience "random genetic drift," according to Wright. Was this "random genetic drift" important to evolution, or merely back and forth variation:

> ... the demonstration that mutant genes may gain high frequencies in a population by the mechanism of genetic drift does indicate the possible role which this mechanism may play in progressive evolution. If, as Wright has postulated, favorable gene combinations form adaptive peaks leading to a better adjustment of the organism to its environment and thus to progressive evolution, genetic drift may play a considerable role in concentrating the gene constituents which enter into the formation of such adaptive peaks, and in increasing the chance of such peaks being formed on a scale large enough that they will not immediately be worn away by the weathering effects of accident when in too low concentrations.
> (Spencer 1947, 109)

Wrightian "random genetic drift" (apparently at every locus), according to Spencer, can help in progressive evolution through Wright's shifting balance theory.

Not only that, "random genetic drift" could also help in speciation:

> Along with adaptive characters there must be evolved incipient isolating mechanisms; it would seem that genetic drift might well concentrate the constituent genes of such isolating mechanisms in sufficient numbers so that they could gain a foothold in some spatially isolated section of the species, an island population. (Spencer 1947, 109).

Ernst Mayr was the editor of *Evolution* and had written similar views about Wright, progressive evolution, and speciation in his 1942 book, but without much evidence. Spencer was providing much more careful field evidence of "random genetic drift" and its effects.

Spencer accepted Wright's version of "random genetic drift": when a population gets small, "random genetic drift" gets larger at every locus on every chromosome. Small winter populations of *Drosophila immigrans* undergo an inbreeding effect, but Spencer said nothing about inbreeding, and followed Wright. Spencer argues that "random genetic drift" had an intimate relationship to differentiation and speciation. Spencer's experiments were devoted to inbreeding, not "random genetic drift" on a chromosome.

Reed and Reed 1948

Sheldon C. Reed and his wife, Elizabeth C. Reed, organized the first experiments on "random genetic drift" in laboratory populations of *Drosophila melanogaster* at Harvard. Wright, Dobzhansky, and Ed Novitsky all read the manuscript. Reed and Reed invented a special arrangement of Dobzhansky's population cages in smaller bottles so that mites and fungi were more controllable. The idea was to choose an X chromosome with M5 (Muller-5) with a long inversion that made both the males with it and homozygous females semi-sterile with poor viability, and also with excellent markers (apricot and Bar). Another stock had a *wmf* combination on the X chromosome (white eyes, miniature wings, forked bristles) with near-normal viability in homozygous state. If M5 was heterozygous with an X chromosome without M5, no crossing over was possible because of the inversion in M5. The heterozygotes were always more viable than the homozygotes leading to the expectation of a balanced equilibrium. The experiment paired heterozygotes (females) with either M5 males or *wmf* males kept together in the bottle for two months, before counting. The heterozygote superiority shone forth, and the balanced polymorphism was maintained by selection as expected. M5/M5 homozygous females could indeed have crossovers but M5 males never seemed to breed.

The Reeds believed their population showed "random genetic drift." They could not distinguish the effects of environmental fluctuations and accidents of sampling, but could see that small populations transferred to new bottles had larger variations in chromosome frequencies than did larger populations. This difference in outcome, they said, stemmed from "random genetic drift":

> The smaller the population at the starting count the greater the change may be at the next count. The fact that this relationship holds is a demonstration of

sampling variability at the population minimum which would result in "genetic drift" except that in our experiment natural selection is too strong to allow much cumulative effect of the sampling variability.
(Reed and Reed 1948, 183-184)

Wright contributed the calculation from the data of the selection rates favoring the heterozygotes, and offered comments about the entire paper. Dobzhansky and his student Ed Novitski also read the paper and offered comments. Reed and Reed had the best and most-informed critics possible.

Reeds's research did not demonstrate Wrightian "random genetic drift" on one chromosome. They created experimental populations by counting between bottles for later populations. They made no distinction between founder populations started in this manner, with inbreeding being prominent, and Wright's "random genetic drift." The Reeds had absorbed Wright's views on "random genetic drift." They should have examined whole chromosome behavior since no recombination was possible (rather, they did have some recombination, but eliminated these bottles from the experiment when it happened). Nor did they distinguish between chromosomal variation and that at two specific loci. Wright, following his own views, could not help being a poor guide to their experimental work.

No Wrightian "random genetic drift" is demonstrated in the Reed and Reed paper, only inbreeding effects, but Reed and Reed never mention it. In the conclusion, they did mention the "accidental variations of the M5 and *wmt* chromosomes," but what relation this bears to Wright's "random sampling of gametes," meaning Mendelian random mating, is unclear. "Random genetic drift" gets in the way of understanding the inbreeding effect. Reed and Reed do not mention inbreeding effects, and invoke a vague "random sampling" that supposedly provides "random genetic drift" from "random sampling of gametes," which is random breeding of gametes as the first state of Mendelism.

Isidore Ludwin 1951

A fellow Harvard student with Reed and Reed, Isidore Ludwin, using the same kind of bottles and experimental setup as Reed and Reed, researched natural selection: "Natural selection in *Drosophila melanogaster* under laboratory conditions" (Ludwin 1951). The aim of the paper was different than the Reed and Reed paper, because natural selection and "random genetic drift" were both involved. Using four different loci at widely different locations on the X-chromosome, Ludwin investigated "the effect of natural selection" and "random genetic drift" upon the frequency of four sex-linked characters either alone or in all possible combinations in highly uniform continuous populations of *D. melanogaster* (Ludwin 1951, 231). The huge difference with the Reed and Reed research on "random genetic drift" was that Ludwin used no inversion, so known map distances between used characters suggest that some crossovers in meiosis were likely. The population sizes were between 200 and

300 "where appreciable random genetic drift might be expected" (Ludwin 1951, 241). Ludwin devoted four of the five paragraphs of the discussion section to "random genetic drift" in his populations, and suggested the ways to detect selection and "random genetic drift."

> The effects of natural selection are recognized by orderly population changes in gene frequency. Such changes may lead to gene loss, fixation, or equilibrium. Whatever the end, it results from consistent orderly change toward the final state. A necessary condition is the size of the effective breeding population be large enough to preclude random gene loss or fixation. Random genetic drift, a different process, is recognized by random variation in gene frequency of unit groups which depart widely from the gene frequency of the complete population. Both natural selection and random genetic drift affect the populations described and though the latter proved a valuable concept, natural selection was more effective in determining gene loss or fixation. (Ludwin 1951, 241)

Wright's "random genetic drift" seemed important in Ludwin's experiments, and he emphasized its importance, but never mentioned inbreeding. He could find "random genetic drift" in only one mutation that he assumed would end with an equilibrium frequency of .5, but actually showed significant deviations from equilibrium.

Ludwin concluded: "This is believed to be due to random genetic drift" (Ludwin 1951, 241).

Ludwin used four visible mutations on the X chromosome, located far apart. Some recombination, even in a small population, would take place between the four markers. Selection rates were high on three of the markers when separated, or in common with other markers, and the fourth, "raspberry," showed some significant variation when alone. The basic problem was monitoring what else on the X chromosome bearing "raspberry" might be influencing the "frequencies of raspberry." By this time, Ludwin had only "raspberry" remaining on the X chromosome. He used the "raspberry" locus as the key to the survival of the whole X chromosome, which changed every generation in meiosis. Assuming that the chromosome had the same selective value as the "raspberry" locus was not proved in this experiment. Meiosis continued but Ludwin paid no attention.

Wright's influence led Ludwin to invoke Wright's views of "random genetic drift" rather than the much clearer understanding of inbreeding effects. What is impressive about Ludwin's research was using four alleles on the X chromosome, and he had meiotic recombination between them. Now we know clearly that this does not spawn "random genetic drift" in the population, but is a result of meiotic recombination.

David Merrell 1953

Neither Reed and Reed nor Ludwin had set up their experiments to keep population sizes really small, but Reed and Reed (1948) pointed out the possibility of using an even smaller bottle size to accomplish that aim. David Merrell, who had conducted a selection experiment for his doctoral research working with Sheldon Reed at Harvard (Merrell 1949), hoped to design a better experiment on "random genetic drift" using the smaller bottles suggested by the Reeds. With a substantive historical introduction that began with the struggle of Fisher, Ford, and Wright on "random genetic drift" in *Panaxia dominula* (Provine 1986a, 420-437), Merrell entered the controversy with an experiment designed to weigh on the side of Wright, showing "random genetic drift" and natural selection at work together.

Using four sex-linked recessives and one autosomal recessive, Merrell took each alone on a chromosome and put into a cage five homozygous mutant males, five wild-type brothers of these five, and five heterozygous sisters. Merrell did not wish to follow Ludwin using all four sex-linked characters on a single chromosome:

> Ludwin (1951) has shown that combinations of linked genes break up rapidly in populations due to crossing over, and each mutant pursues the same course in these populations that it does when alone in competition with its wild type allele. Therefore, it seemed unlikely that any combination of other genes

might persist in these populations which would noticeably change the results from those obtained with flies which were more nearly isogenic. Furthermore, the selective disadvantage of the mutants was so great (except, perhaps, for forked) as to overshadow any minor differences existing at other loci. (Merrell 1953, 96)

Merrell dismissed all differences between chromosomes as being so small in selective value in comparison with the mutant at the locus that all chromosomes could be treated equally in the experiment. The "random genetic drift" in this experiment had turned back into the Wright model; the locus was the same thing as a chromosome. Selection was chromosome selection, and "random genetic drift" of whole chromosomes, better described as inbreeding in chromosomes in a small population. Of fifty seven populations, twenty seven went extinct, according to Merrell, "at random, because no differences were noted between the characteristics of the surviving populations and those which were lost" (Merrell 1953, 96).

Merrell's data showed clearly that all the mutant recessives tended toward elimination in the cages, thus exhibiting the effects of adverse selection. In no case during the experiment did a population become fixed in the recessive, but many lost the recessive, and Merrell concluded that "random loss of the recessive genes has evidently occurred in several cases" (Merrell 1953, 100). When population sizes grew small, he explained:

> ... genetic drift now reinforces the action of selection since it generally tends to reduce still further the frequency of the less frequent allele in a population. Thus, on the average, homozygosity for the more favored allele will be reached more rapidly in small populations than in large because of the combined effects of selection and genetic drift. (Merrell 1953, 100)

He concluded that "large fluctuations in gene frequency occurred due to random genetic drift, in some cases leading to loss of the recessive gene" (Merrell 1953, 100). All of Merrell's talk about mutants and wild types at a locus was really talk about whole chromosomes. His experiment was inbreeding with loss of whole chromosomes.

Merrell's mutants survive poorly against the wild type, and do not disappear or get scarce gradually. "Most striking, perhaps, are the large fluctuations in gene frequency which occur from one count to the next in so many cases" (Merrell 1953, 98), leaving "large fluctuations in chromosomes." Merrell had turned inbreeding into "random genetic drift" of chromosomes. His results are better explained by inbreeding.

He also sounded a note of caution about using genes under adverse selection to test for "random genetic drift," since detecting the precise role of "random genetic drift" was impossible if selection was involved. "Unless some alleles can be identified which are neutral under all conditions, changes in gene frequency in small populations must be interpreted in terms of both selection and random genetic drift," as in the case of this experiment.

Tim Prout 1954

A far more careful and substantial experiment on "random genetic drift" came from the Dobzhansky lab at Columbia. Timothy Prout, a doctoral student with Dobzhansky, did his thesis research on "random genetic drift" in three irradiated experimental populations of *Drosophila melanogaster*, published as his doctoral thesis in 1953. Prout described the general process he used to show "random genetic drift" in recessive lethal alleles. He focused upon loci with a pair of alleles:

> If all these alleles are subject to the same evolutionary forces (mutation, selection, migration, etc.) then the variance of allele frequencies may be taken as a measure of genetic drift. In other words, genetic drift will tend to cause the alleles at some loci to be represented more times than the average and some less than the average.
>
> It is this model which will be used in the present investigation of the effects of genetic drift on the behavior of recessive lethal alleles at various loci on the second chromosome of *Drosophila melanogaster*. (Prout 1954, 529)

Prout sampled three interesting populations. Population #3 was a large population (about 10,000 individuals) with no radiation. Population #6 was the same size, but exposed to radiation that produces mutations. Population #5 was a decidedly smaller population (about 1000 individuals) but also exposed to radiation. The radiation was known to produce recessive lethals, and Dobzhansky, Wright, and Bruce Wallace (in whose lab these populations resided) had worked out breeding experiments to reveal the allelism of these lethals.

The experiment showed that the allele frequencies of lethals was constant in large populations, but varied up and down in the small population #5; Prout concluded that "random genetic drift" was causing these variations in the frequencies of lethals, and he offered the standard explanation, beginning with effective population size:

> In a finite population, even with random mating and randomly varying family size, a variance of gene frequencies is expected to develop. This will occur because of the sampling errors involved in the choice of a finite number of gametes from the gene pool of one generation to produce the zygotes of the next. This random drift variance can be further increased by mating of relatives, or by increased variation of family sizes (that is, where a large number of individuals leave no offspring, or a few individuals have a large number of offspring, or both). If one or both of these factors are in operation, then it is said that the

effective size is smaller than its absolute size. (Prout 1954, 540)

Prout used the "gene pool" (Chapter Six) with his variations. With some assumptions, Prout was able to calculate the effective size of the small population, with a combined estimate of N = 256, as opposed to about 1000 individuals in the absolute size. Such a small size should immediately raise the problem of "random genetic drift" in this population. "The data," said Prout, "have provided strong evidence of the operation of genetic drift" (Prout 1954, 543).

According to Wright, most small populations went extinct. Prout, too, was worried about the evolutionary importance of "random genetic drift," despite having just shown its existence.

The study of lethal producing loci, then, has served as an effective method for detecting genetic drift. It is doubtful, however, that drifting lethal frequencies can have any great effect on the evolutionary history of a population. It is obvious, however, that if there is drift variance in lethal frequencies there must be drift variance of the frequencies of less deleterious alleles. In fact selection is operating more severely against drift variance of lethal alleles than it is against the variance of less deleterious alleles. In the small population then there must be an even greater non-adaptive variance of the alleles which are expected to be of greater evolutionary significance than lethal alleles.

> At this stage in the experimental investigation of genetic drift is not of first importance that the material used for detecting drift be of evolutionary significance. (Prout 1954, 543)

Mendelian inheritance produced apparent "random genetic drift," so he believed Wright's views. Having evolutionary significance, however, was a separate question.

Prout certainly has accepted the Wright model fueled by Mendelian inheritance, and thought that random sampling of the germ cells provides "random genetic drift" at all loci on the chromosome, using the "gene pool." What Prout has actually shown in his paper on "random genetic drift" in *Drosophila melanogaster*, and what he has assumed from Wright, fit together well.

Prout spoke about loci and lethals at the locus. His Wright model inevitably was based upon whole chromosomes. He did not need "random genetic drift" to explain what he was seeing. All he needed was inbreeding effect, which loses chromosomes in an effective population size of 256 individuals.

Ludwin brought out the real problems of talking about "random genetic drift" with four loci, widely spaced, on the X chromosome, pointing to meiotic recombination. Both Merrell and Prout went back to the simplest Wright model, treated the whole chromosome as one locus, and clamored about having demonstrated "random genetic drift" at a locus in a small population. Ludwin had used inbreeding and taken the effects as "random genetic drift" and so did Prout.

Random drift in human populations, 1950-1954

In late 1950, Curt Stern, a well-known *Drosophila* geneticist, wrote his textbook, *Principles of Human Genetics*. His book held a special place at the time, not because it was the only one available, but the one written by a modern, highly respected professional geneticist, whose purpose was not simply medical genetics, but human genetics. Stern devoted seven pages to "random genetic drift" in human populations. He explained the concept in some detail and then suggested its application: whenever human population sizes were small, "random genetic drift" was also large:

> Drift in its various aspects probably accounts for much of the polytypy of man. While the relative homogeneity of racial groups in respect to some traits is probably the result of selection for certain genotypes, uniformity for other racial traits may have resulted from loss and fixation caused by drift. Drift, in its less extreme form of shifting allelic ratios exclusive of loss and fixation, is also probably responsible for many of the polymorphic differences of different races. (Stern 1950, 597-598)

As examples, Stern used variation in some blood groups and alleles determining diseases, though he warned about indiscriminate use of "random genetic drift" to explain all such genetic differences between human populations.

The same year, 1950, Stern served as chairman of a session of the Cold Spring Harbor Symposium, that year devoted to the topic, "Origin and Evolution of Man." The longest paper in the published version was a huge article (at fifty five pages, more than three times as large as any other paper) by Joseph B. Birdsell discussing the role of "random genetic drift" in determining the differences between populations, usually small, of native Australians. His examples included blood group frequencies, stature, total facial height, and the distribution of supernumerary fourth molar teeth. In each, he could find no significant selective factors and concluded that observed differences between populations was produced by "random genetic drift." Other differences, such as tawny hair, he attributed to selection. His argument for "random genetic drift" takes up most of Birdsell's article.

He added, however, a caveat: "*Thus drift, at best, can be no more than suggested in terms of broad probabilities.*" (Birdsell 1950, 262). I suspect, but cannot prove, that Birdsell added this caveat because of the long critical remark that Arthur Cain gave in response to Birdsell at the meeting.

> "There is a very real danger that when no obvious selective influences can be found to explain a particular variation in gene frequency, the investigator will conclude that therefore drift must be acting"

(Cain comment, in Birdsell 1950, 312). Citing Wright, Birdsell responded that emphasis upon the single factor of selection was a poor way to analyze genetic differences between small populations of Australians. Dobzhansky chimed in with his support for "random genetic drift" as one important factor in addition to selection in the populations studied by Birdsell.

Bentley Glass, present for the Birdsell presentation and discussion, published a second, and more influential with geneticists, study of "random genetic drift" in human populations (Glass *et al.* 1952). The introductory discussion related the study to the arguments of Wright with Fisher, Ford, and Phillip Sheppard over "random genetic drift." Glass continued:

> The interesting discussion on this point between Cain, Birdsell, and Dobzhansky (Birdsell 1950) has served to emphasize that random genetic drift and selection are not alternative explanations but are factors which may often act in conjunction, as Sewall Wright has constantly maintained. Nevertheless, the conclusion that random genetic drift is at all responsible for the observed differences in gene frequencies in the Australian tribes is rendered uncertain by the possibility that selection might operate in these circumstances and be the chief or sole cause of the genetic differences. (Glass 1953, 146)

The purpose of this study was first to avoid the pitfalls of the Birdsell work, and to provide robust evidence of "random genetic drift" in human populations.

Glass chose a religious isolate, "Dunkers" (Old German Baptist Brethren), living in a close community in Pennsylvania with a measured population size of 298 individuals, and an equally definite breeding size of 90 parents ($N = 90$), with historical evidence to support $N = {<}100$ since 1895. The control groups were the much larger Dunker population in Duisberg, Germany, and select populations from the USA (for example, Baltimore white high school students). The population had 10% to 15% emigration and immigration each generation, thus maintaining a constant population size. For his analysis, he chose seven loci: ABO blood groups, MN blood groups, Rh blood groups, Middigital hair, Distal hyperextensibility of the thumb, ear lobes, and handedness.

At five of the loci, he found "evidence" of "random genetic drift," which came from three causes:

> The random genetic drift resulting from accidents of sampling [random sampling of gametes] because of the small effective size of the population (N) appears therefore in this community to be compounded of three elements: (a) the accidental composition of the original community in Franklin County: (b) the effects in each generation of the random sampling of gametes in a very small population; and c) the chance exclusion of genes borne by those who leave the community. (Glass *et al.* 1952, 151)

Many people who mimic Wright use this as their cause: accidents of sampling gametes, which produces no "random genetic drift," but is inbreeding.

Glass dismissed the possibility of selection of some sort producing the observed differences in variation on his traits of genes:

> Considering the fact that the traits studied are in all cases common and considering also the homogeneity of environment of the individuals comprising the isolate and those in the surrounding American population, it seems unnecessary to attribute these divergences to some hypothetical influence of selection. In fact, the very loci where selection most demonstrably operates, namely, the Rh and handedness loci, are those where no appreciable drift appears to have occurred. (Glass *et al.* 1952, 158)

Despite dismissing the importance of selection as an explanatory factor in observed differences in gene frequency, Glass added: "The conclusion to be derived from the study, that random genetic drift can in fact determine gene frequencies to considerable extent in small human isolates, is still a tentative one" (Glass *et al.* 1952, 159).

With time, Glass became more certain about his study. In *The Scientific American*, he wrote: "there was no explanation for these novel combinations of hereditary features except the supposition that random genetic drift has been at work" (Glass 1953, 80). In a summary article in *Annual Review of Genetics*, "Genetic changes in human populations, especially those due to gene flow and random genetic drift" (Glass 1954), he described the study as "conclusive evidence of the operation of random genetic drift" (Glass 1954, 135).

Geneticists and evolutionists generally agreed with Glass's work, but actual evidence is but inbreeding. The Glass study was referred to constantly for the next 55 years in books on human genetics or human evolution. Dobzhansky, in his summary book on genetics and evolution, *Genetics of the Evolutionary Process*, called the study "elegant" (Dobzhansky 1970, 246).

Stern, Birdsell, and Glass followed Wright who had by 1954 become the central authority in evolutionary biology on population size. The connection of "random genetic drift" to small populations had become truth in evolutionary biology. Glass never mentioned inbreeding, the cause of what he observed in humans.

New experiments: Wright and Kerr, Kerr and Wright 1954

Warwick Estevam Kerr came from Brazil, where he earned his doctorate, to the USA in 1951 as Visiting Professor at University of California at Davis and then in 1952 as Visiting Professor at Columbia University working with Dobzhansky. During his stay at Columbia, Kerr decided to provide, experimentally, such decisive proof of "random genetic drift" that the issue would be settled. He received help from J. F. Crow at the University of Wisconsin in Madison for experimental specimens and advice on methodology.

Dobzhansky, and even his mathematical advisor at the time, Howard Levene, and Jim Crow thought that his experimental data needed quantitative help from one person: Sewall Wright himself. Who would be better qualified to measure "random genetic drift," and sort out selective factors from random drift? Wright had withdrawn himself from Dobzhansky's *Genetics of Natural Populations* series of papers after five collaborations, the last in 1947. Wright was preparing to leave the University of Chicago for the University of Wisconsin in 1954, his last year at Chicago. Collaborating with Warwick Kerr must have been a tough decision for Wright. From where else would come such resounding support for "random genetic drift"? Even though Wright had left behind his assertions about how "random genetic drift" led to non-adaptive differences in closely related species, he nevertheless strongly supported "random genetic drift" in large sub-divided populations, the key to his shifting balance theory. "Random genetic drift," Wright reasoned, provided the kaleidoscopic shifting of gene frequencies at every locus on every chromosome that provided his shifting balance theory with ample genetic variation. Wright decided to collaborate with Kerr, who would do the experimental work, and Wright the quantitative analysis.

The experimental idea behind Kerr's experimental work was direct and simple and extended to three series of *Drosophila melanogaster*:

> About 120 lines were started in each [series]. In the first series, the sex-linked mutation forked (f) competed with its type allele. Four females (1f/f, 2 f/+, 1+/+) and 4 males (2f/0; 2+/0) were put in each vial. The second series involved the sex-linked semidominant mutation Bar (B) and its type allele.

Each initial vial contained 4 B/+ females and 4 males (2B/0, 2 +/0). The third series was with the autosomal alleles aristapedia, ss^a, and spineless, ss; which produce a heterozygote that is close to type. Each initial vial contained 4 ss^a/ss females and 4 ss^a/ss males. The alleles were thus equally frequent at the beginning of each experiment.

The cultures were allowed to develop until about 2 to 4 days after the offspring began to emerge. The flies hatched up to this point were discarded (in an evening). The flies which appeared next morning (if enough had emerged) were etherized and from among them 4 females and 4 males were taken at random and served as progenitors of the following generation. The etherized flies were put on a porcelain plate and the first 4 males and first 4 females that happened to be closest to the right end of the plate were the flies taken. It was often, however, necessary to wait to the second and sometimes to the third day to obtain 4 of each sex. This procedure was repeated in every following generation in every line. The first series was carried 16 generations, the second 10 and the third 9 generations. (Kerr and Wright, 1954a, 172-173)

Population size was large, and the sub-populations small. If a population became fixed or could not produce 4 females and 4 males, the lines were ended. For forked, 96 lines were counted. Wild type was fixed in 41 lines and forked in 29 lines, and 26 lines were still heterozygous after 16 generations. These results, the authors claimed, were in agreement with the work of Ludwin and Merrell on random drift in *Drosophila melanogaster* using forked.

The discussion section is crucial for understanding why Wright participated in the project with Kerr. I believe this section was written or edited by Wright:

> Populations of 4 males and 4 females per generation are so exceedingly small that experiments such as the present may seem to have no implications for evolution in nature. It must be borne in mind, however, that changes in the underlying multifactorial genetic structure of species probably occur so slowly that an appreciable change in a thousand generations must be considered as an explosively rapid process.
>
> Study in the laboratory of the factors that can contribute to such change is practicable only by stepping up the rates by at least one hundred fold. Thus the interaction between a weak selective advantage of one isoallele over another and a slight random drift, due to inbreeding, can be simulated by using alleles with selective differentials of ten percent or more instead of perhaps only one tenth of a percent or even one hundredth of a percent in populations of only one percent of the size of a typical natural deme.

> In the case of forked, the selective differential is clearly much less than ten percent so that the results of the present paper illustrate random drift from inbreeding in an almost pure form. (Kerr and Wright 1954a, 177)

Wright had given up "random genetic drift" as the cause of species differences, and more certainly, differences between genera. He could see his shifting balance theory in progressive evolution, speciation, and in the fossil record. "Random genetic drift" still was a crucial component of the shifting balance theory. "Random genetic drift" was a component to what Wright calls "the multifactorial genetic structure" of a population. The experiment with Kerr really was on inbreeding. By 1954, Wright had completely confused inbreeding with "random genetic drift." Even Wright bills this experiment as inbreeding, so the confusion of inbreeding and "random genetic drift" was complete in Wright by 1954.

In series two, Kerr used the sex-linked semi-dominant mutation Bar (B) and its type allele. Bar clearly did not survive well: 95 lines fixed with the wild type in 10 generations, Bar was fixed in 3 lines, and only 10 lines were left unfixed. Their conclusion was stated at the beginning: "selection strongly favors type but . . . Bar can occasionally drift into fixation in spite of this" (Wright and Kerr 1954a, 225).

In series three, Kerr used the autosomal alleles aristapedia, ss[a], and spineless, ss. The heterozygote was far more fit than either homozygote. Spineless "drifted" into fixation in eight lines, after 9 generations. The discussion section of this paper is the heart of Wright's view of Kerr's entire series of experiments:

> In natural panmictic populations of ordinary size, genes subject to such enormous selection, as that against Bar in the second paper of this series, would be eliminated too rapidly for random drift from inbreeding to be of appreciable importance. The same is probably true of forked (in the first paper) even though the selection against it was hardly detectible in the small experimental populations. In the case of aristapedia and spineless, the enormous selection against both homozygotes would keep a large very close to equilibrium point (about 40% ss_a). We may, however, consider the experimental populations as models, on exaggerated scales with respect to the effects of inbreeding and, in two cases, selection, of situations that might occur in nature with respect to a pair of isoalleles or of alleles with primary effects on multifactorial quantitative variability.
>
> In arrays of such populations with effective size one thousand times as great as in the experimental populations, the random drift and ultimate fixation due to inbreeding would occur at only one-thousandth of the rates observed here. The forms of the distributions after approximate stability has been reached (in one thousand times the observed

periods) would, however, be nearly the same as observed if selective disadvantage in all cases is only one-tenth of a percent of those in the experiments. Such coefficients are reasonable enough for isoalleles or alleles with primary effects on quantitative variability.

The experiments with forked may be considered as giving a model of the inbreeding effect in almost pure form. Those with Bar give a model of the case in which an allele tends toward fixation but not so decisively as to prevent the unfavorable allele from occasionally drifting into fixation against the pressure of selection. Those with aristapedia and spineless illustrate an approach to equilibrium between alleles because of selection against both homozygotes, but balancing of these selection pressures against random drift from inbreeding in such a way that gene frequency varies almost from one extreme to the other among the populations and one at least of the alleles may occasionally drift into fixation. (Kerr and Wright, 1954b, 300-301)

Wright's direct statement above, "random drift from inbreeding," is a contradiction of our understanding now. Wright's conclusion, that selection and "random genetic drift" will inevitably interact in every real natural population with different effects depending upon the population and genes examined, was almost lost to his reading public:

geneticists and evolutionists. Their question was, did these experiments demonstrate that "random genetic drift" really works? Kerr and Wright said yes, "random genetic drift" works, and from my perspective, not at all. For decades after this, including the present, evolutionists or geneticists quote these three papers by Kerr and Wright as demonstrating the efficacy of "random genetic drift" in small populations.

In my opinion, what Kerr and Wright showed was that intense inbreeding shows the results of inbreeding, not "random genetic drift." The genes used were simply to mark the chromosomes and to add problems with selection. All the mechanics were at the chromosomal level. In the early time before 1924, Wright, East, and Jones would have used only inbreeding, and none would invoke "random genetic drift" as demonstrated by Fisher.

Wright is clear throughout these papers with Kerr that inbreeding causes "random genetic drift." Wright said in other papers before and after the papers with Kerr that "random sampling of gametes" caused "random genetic drift." Neither of these causes can produce Wright's view of "random genetic drift" on every locus on every chromosome. All that Wright and Kerr showed in the experiments is inbreeding effects expected with whole chromosomes.

Crow and Morton 1955

In 1955, a paper on the theory of "random genetic drift" appeared in *Evolution*: "Measurement of Gene Frequency Drift in Small

Populations" (Crow, J. F. and Morton, N. 1955). Crow and Morton submitted to *Evolution* because they wished to reach evolutionists. Using a one-locus model and the "gene pool," they demonstrated how to determine effective population size N_e, a long-standing problem in population genetics. They explained the variance of an allele at this locus $V_{\delta q}$ (giving an estimate of the dispersive factors such as "random genetic drift"), and they proposed a new index of variability, which appeared in the equation defining N_e. Also appearing in this equation was F, Wright's coefficient of inbreeding. After this theory, they described an experiment on "random genetic drift" in *Drosophila melanogaster*, with "gene pools."

Crow and Morton knew from Wright that "random genetic drift" was effective in low population numbers, or size (N_e), and the portion of the variance in gene frequency changed in one generation from dispersive factors ($V_{\delta q}$) due to sampling variance from a finite population. N_e was a tough problem to calculate directly except in unusual circumstances, and the same was true of $V_{\delta q}$.

By using "gene pools" in addition to one-locus models, Crow and Morton were able to derive these quantities and then found them by direct census methods and showed how to use their formulas in a laboratory population of *Drosophila melanogaster*. The "gene pools" enabled them to work on this problem. Note that all Mendelian processes were allowed to choose from the "gene pool," or add back to it. I will return to "gene pools in Chapter Six.

Crow and Morton did a "random genetic drift" experiment and calculated N_e and $V_{\delta q}$. This experiment was an inbreeding experiment. They were talking about whole chromosomes by referring to the mutants on the chromosome. No "random genetic drift" on a chromosome occurred in this experiment. They used an advantaged method: using a "gene pool." You can take anything from it, and put anything back into it. Gene pools are not real biological entities.

Buri 1956

Peter Buri was the last of Wright's students at the University of Chicago before Wright went to the University of Wisconsin in Madison in 1955. Buri followed the experimental procedures used by Kerr and Wright. Buri described his experimental population: "The form is that of a fairly large population divided into a number of very small completely isolated units." (Buri 1956, 367). He used far more small populations and larger populations than Kerr. With more of both very small and larger populations in his experiment, Buri should have more decisive results than Kerr and Wright.

Buri chose to work with the brown locus bw and another strain of bw^{75} that Herman M. Slatis developed in the early 1950s. Both came from stocks highly inbred by crossing both and mostly brother-sister mating, a technique that gave both the bw and bw^{75} the same isogenic (almost homozygotic) stock on the 1st and 3rd chromosomes; the second chromosome was isogenic except for very near the brown locus. The stocks had at least 40 generations of inbreeding and crossing.

Each stock was founded with 16 bw/bw^{75} hybrids at the brown locus, 8 male and 8 female. One series used 35cc. bottles and the other 60cc. bottles. As a selection control before his experiment, Buri turned to selection in both bottles and population cages. He found no selective differences between the mutants bw and bw^{75}. When Buri turned to "random genetic drift," he also found again that selective differences between bw and bw^{75} using 35cc. or 60cc. bottles showed no differences. Thus selection (using this common isogenetic genome of both stocks) showed no differences for the "random genetic drift" experiment. He could just ignore selection and concentrate upon "random genetic drift."

In the section of his paper devoted to "random genetic drift," Buri used the paper of Crow and Morton (1955) discussed above, and calculated N_e and found values different in the 35cc. and 60cc. bottles. He nevertheless concluded from his results in this "random genetic drift": "the distributions of the series I [35cc.] and Series II [60cc.] are statistically very similar" (Buri 1956, 385). Even though Buri used different population sizes, he found no evidence of extra "random genetic drift" in the smaller populations. His only way to calculate the effect of "random genetic drift" was to argue that selection counted for nothing, and "gene frequency changes could safely be attributed wholly to accidents of sampling." Buri meant that accidents of sampling gametes was the source of "random genetic drift" in these populations.

Neither Wright nor Buri considered this experiment to be insightful into "random genetic drift." The experiment was entirely an inbreeding experiment using whole chromosomes. Any random drift would have to be of whole chromosomes. This paper shares all the problems of Kerr and Wright, and Wright and Kerr (1954), and demonstrates no "random genetic drift" on a chromosome.

Buri could not know at this time that the experiments of Kerr and Wright would soon kill off all other experiments on "random genetic drift." Experts cited proof of "random genetic drift" with Kerr and Wright, Wright and Kerr. The experiments on "random genetic drift" addressed in this chapter have, with the exception of Kerr and Wright, and Bentley Glass, largely disappeared from the genetics and evolutionary literature.

Dobzhansky and Pavlovsky 1957: gene pools and "random genetic drift"

Dobzhansky often used the term "gene pool" in the early 1950s (Dobzhansky 1951a, 1951b); we have already seen how important this invocation was for one-locus models of Mendelism in the work of Crow and many others who used it to great advantage in Mendelian population genetics. Ernst Mayr also found this idea enormously appealing entirely apart from Mendelian population genetics because he used it for speciation and his "genetic revolutions" (Mayr 1954, Provine 2006). We shall see in this Dobzhansky and Pavlovsky paper how important Mayr's 1954 paper was to them.

Dobzhansky knew this 1957 paper was coming out after a spate of papers on "random genetic drift" tested in the laboratory, and he also suspected this would be the last of this series of papers on "random genetic drift." His aim was to demonstrate both "random genetic drift" and selection. Dobzhansky put the entire series into historical perspective, pushed his view that interaction of "random genetic drift" with natural selection was the key to understanding both, and talked about the ultimate addition of "random genetic drift" to evolutionary biology as a crucial factor in the process of adaptive evolution and speciation.

Dobzhansky and Pavlovsky began their paper: "The role of random genetic drift in the evolutionary process has, for about two decades, been one of the most controversial issues in population genetics" (Dobzhansky and Pavlovsky 1957, 311). They were careful about using "random genetic drift" as whatever is leftover from selection.

Dobzhansky used old populations in his lab. Two inversions in *Drosophila pseudoobscura* came from Texas (Pikes Peak, PP) and from California (Arrowhead, AR). The populations differed extensively, not just from different inversions on chromosome 3. Hybrids give rise to many meiotic recombinations of chromosomes that show different results might be expected from one hybrid population to another. Dobzhansky tried to put selection in perspective by conducting two selection experiments from hybrids of PP and AH. Starting at 50% of each inversion, in 45 generations, the two populations both showed decrease in PP coming to rest at 25% PP in one population cage and 41.7% in the other population. This, said Dobzhansky, showed that the results were clearly different, but "not especially great" (Dobzhansky and Pavlovsky 1957, 313). Selection could give a wide array of results regarding the two inversions, but perhaps due to other genetic differences rather than the differences of the inversions.

Dobzhansky started 20 more population cages by putting 4000 F_2 hybrids into 10 containers, and filled the other 10 cages with F_2 hybrids, 10 females and 10 males. After one generation in the cages, all were filled and treated alike. Over five months, the distribution of offspring was tested, and found to be very similar in averages and variances between one population and another. Over the following 13 months, the 10 populations started with 4000 hybrids and the 10 populations started with 20 hybrids had statistically the same percentage of PP chromosomes averaged over all 10 populations. The cages started with 20 individuals changed in this interval showing much greater variance between its 10 populations than that of the 10 populations started with 4000 hybrids. Dobzhansky found no effects on correlations between characters in the populations during the five month test, but found some correlations between characters after 18 months.

Why did the differences between populations started with 20 hybrids increase significantly between 5 months and 18 months of the experiment? Dobzhansky did suppose an answer:

> . . . the selective fates of the chromosomal gene arrangements become dependent upon the poly-genic genetic background, which is highly complex and variable because of the gene recombination that is

> bound to occur in populations descended from race hybrids. Here random drift becomes operative and important. It becomes important despite the populations being small only at the beginning of the experiments, because the foundation stocks in some populations consisted of small numbers of individuals. Thereafter, all the populations expand to equal sizes, fluctuating roughly between 1000 and 4000 adult individuals. Such populations can be regarded as small only in relation to the number of gene recombinations which are possible in populations of hybrid populations. (Dobzhansky and Pavlovsky 1957, 316)

Dobzhansky is confused about inbreeding effects and "random genetic drift." Dobzhansky chose these 20 individuals from different lines of his original stocks of PP and AR, to avoid inbreeding effects. He maximized the amount of variation in these 20 flies, and expected for one generation of small size to produce a lot of "random genetic drift" for the five generations of offspring. His prediction was poor in the one generation possible.

Dobzhansky now introduces the "gene pool" to clarify his position on "random genetic drift." He dismissed tracing the fates of individual genes as important to evolution:

> ... it becomes indispensable to consider not only the destinies of single genes but also of the integrated genotypes and finally of the gene pool of Mendelian populations. In our experiments, the foundation stock

of the populations consisted of F_2 hybrids between rather remote geographical races; a highly variable gene pool arose owing to the hybridization; random drift caused different segments of this gene pool to be included in the foundation stocks of each population, especially in the small ones; natural selection then produced divergent results in different populations, especially again amongst the small ones. (Dobzhansky and Pavlovsky 1957, 317)

The "gene pool" is a huge abstraction, and useless in this paragraph and in this paper. Dobzhansky made a good point. Chromosomal survival is the key to this experiment, not gene survival from "random genetic drift." The number of chromosomes is vastly less in those populations founded by 20 hybrids than in one started by 4000 hybrids. This can have an effect for the duration of the experiment. Such a difference was invisible after 5 generations, and some appeared in 18 generations, a complete mystery.

Mayr (Provine 2005) had no theory of speciation, because he had no theory of the rise of isolating factors. Perhaps his most influential paper was his 1954 paper on "genetic revolutions" and speciation. I have argued that the genetics in this paper were drawn mostly from Bruce Wallace, a student of Dobzhansky. Mayr's genetic revolutions were confusing to Wright, who understood that one generation of a small population would not offer much for adaptive change or speciation. Dobzhansky cited the crucial passage from Mayr 1954:

> Isolating a few individuals (the 'founders') from a variable population which is situated in the midst of a stream of genes which flows ceaselessly though every widespread species will produce a sudden change of the genetic environment of most loci. This change, in fact, is the most drastic genetic change (except for polyploidy and hybridization) which may occur in a population, since it may affect all loci at once. Indeed, it may have the character of veritable 'genetic revolution.' Furthermore, this 'genetic revolution,' released by the isolation of the founder population, may well have the character of a chain reaction. Changes in any locus will in turn affect the selective values at many other loci, until finally the system has reached a new state of equilibrium. (Mayr 1954, 170).

To which Dobzhansky added: "The outcome of our experiments described above may, in a sense, be regarded as experimental verification of Mayr's comment" (Dobzhansky and Pavlovsky 1957, 318). Mayr's 1954 view of speciation was a thesis that he hoped would be translated into biology. Dobzhansky adopted Mayr's thesis on speciation and tried to give proof. He did not succeed.

The experiments of Dobzhansky and Pavlovsky demonstrated no genetic revolution in their populations of *Drosophila pseudoobscura* started with 20 flies. They showed, just possibly, the effects of one generation of inbreeding, but the effects waited not 5 but 18 generations to appear. They demonstrated no "random genetic drift" on a chromosome. Dobzhansky and Pavlovsky confused "random genetic drift" with inbreeding. Dobzhansky cannot demonstrate a "genetic revolution" since Mayr never explains what is such a revolution (Provine 2005).

Conclusion

The decade of 1940 to 1957 saw a host of papers on "random genetic drift." Some were experiments to demonstrate "random genetic drift" in *Drosophila melanogaster* or Dobzhansky's work on *Drosophila pseudoobscura*. Shull inferred "random genetic drift" in human heredity experiments. The last two papers from Wright's student Buri in 1956, and Dobzhansky and Pavlovsky in 1957, were the last "random genetic drift" experiments. I once asked James F. Crow what response he and Morton had to their 1955 paper; did people appreciate demonstration of "random genetic drift"? He answered no, that in 1955, the demonstration of "random genetic drift" was the demonstration of the obvious.

These experiments removed all interest for more than two decades in work on "random genetic drift," which was over-demonstrated. These experiments, especially those of Wright and Kerr, and Kerr and Wright, were cited over and over for evidence of "random genetic drift." Shull's work on "random genetic drift" in humans was also widely used in medical education.

This chapter declares that not one of these papers on "random genetic drift" showed "random genetic drift." Ludwig's paper was the best one of the lot in conception, but he did not demonstrate "random genetic drift" either, or specify the real origin of the changed chromosomes from inbreeding and recombination of chromosomes in meiosis.

The set of papers demonstrate a problem in "random genetic drift" important to my argument in this book. Understanding inbreeding, and understanding its differences from "random genetic drift" in small or large populations is a fundamental problem in genetics. All the experiments showed the effects of inbreeding, and no "random genetic drift." Inbreeding has a real, biological effect. Sewall Wright understood this fact probably better than anyone. Fisher, Haldane, and Wright then confused inbreeding with "random genetic drift."

Chapter Five

Kimura and Ohta

For many years I had great hopes of writing a history of the neutral theories of evolution. My first visit to Japan was in April 1988 to begin my work on the neutral theory. This chapter shows that Motoo Kimura's views on both evolution now and in the neutral theory are based upon Fisher, Wright, and Haldane, and have the same problems. Kimura, Tomoko Ohta, Jack King, and Tom Jukes, of whom only Ohta still lives, and all who work in the neutral theory, are subject to these comments below.

The models of "random genetic drift" used by the young Kimura were precisely those I find in the models of Fisher, Wright, and Haldane. Kimura was their product. He excelled in moving the Kolmogorov mathematical equations into population genetics. "Random genetic drift" is the same in Fisher, Wright, and Haldane as in the neutral theories, including the work of Kolmogorov. I will talk directly about implications of "random genetic drift" and the neutral theories of molecular evolution in this chapter.

Fisher, Haldane, and Wright were unable to meld their knowledge of chromosomes with their equations from statistical physics. They were unable to understand the differences between inbreeding and "random genetic drift" in small populations. We shall see later in this chapter how Kimura, and later Ohta, dealt with Fisher, Haldane, and Wright and their views of "random genetic drift." Kimura did not challenge his colleagues who founded population genetics, but he tried to extend their views by using a different statistical approach by adding to Fisher and Wright. He invented his neutral theory of evolution in 1968, to which Ohta added her slightly deleterious neutral theory in 1973, or later, her nearly neutral theory of evolution.

Kimura and his rise in population genetics

As a student in Eighth National High School in Nagoya, Japan, Kimura became deeply interested in the chromosomes of lilies in the class of his biology Professor M. Kumazawa, whose doctorate was in morphology and cytology. Kumazawa was devoted to his students. Motoo Stimuzo came to Kumazawa in 1942 to ask for chromosome comparisons between his different appearing "species" of lilies. Kumazawa set Kimura to work on this problem. Stimuzo wrote about this collaboration in 1984:

> With his enthusiastic effort and concentration, Mr. Kimura succeeded in clarifying the chromosome morphology of 13 species, 3 varieties, and 16 horticultural varieties among the lilies that I collected. This work was published in a series of three scientific papers (1946, 1947, 1948) authored by Kumazawa and Kimura immediately after the war. These papers were written by Kimura, but, as Professor Kumazawa later used to recount, he made himself the senior author because no reputable science journal would

> accept papers of a high school student. In my opinion, these papers were epoch-making in the field of lily studies. (Stimuzo 1984, 17)

Kimura majored in cytology at the Imperial Kyoto University and his first job was in the Department of Cytology in the National Institute of Genetics in Mishima, Japan. Kimura was an expert in chromosome biology and continued to work on the cytology of orchids all his life except the last year or so when his arms were paralyzed from ALS. He was more of a cytologist than anyone in population genetics, including Fisher, Haldane, or Wright.

While a student in cytology, Kimura became very interested in population genetics. He had no copy machine and for years, Kimura copied out the fundamental papers of population genetics. These papers are extraordinary because they had not only the text, but Kimura performed all calculations left out by Fisher, Haldane, and Wright. Some papers of Wright I understood best from Kimura's transcription where he derived everything. He was an expert on the fundamental papers of population genetics because he worked out every equation and result.

Kimura's early papers on population genetics

Kimura's early papers on population genetics, crucial for understanding his work, were published in the *Annual Report*, the official yearly publication of the National Institute of Genetics. Issue No. 1 covered the years 1949-1950, and was published in English (a fuller edition in Japanese was also available) in 1951. Except for one paper, two pages long, on the *"Stepping-Stone"* model in No. 3 (1952), none of these papers were published with Kimura's collected papers (Takahata, ed. 1994). We must go back to these papers and understand Kimura's aim in the first three volumes where his papers were published in English.

The titles of his papers in 1949-1950, Annual Report No. 1, were:

1. *Effect of Random Fluctuations of Selective Value on the Distributions of Gene Frequencies in Natural Populations*
2. *Physiological Theory of Polygenic Actions*
3. *Recombination of Chromosome Segments though Continued Self-fertilization*
4. *Studies on the Process of Chromosome Substitutions*
5. *Theoretical Relations among Map-distance, Recombination Value and Coincidence Ratio* (Kimura 1951)

Perhaps no one but Kimura in 1951 could have a more wide-ranging combination of topics in his first two years of publishing. He was publishing in Japanese journals, also. The distribution of the first three volumes of the Annual Report from NIG in English were crucial for Kimura's acceptance at Iowa State, and his studies that followed at the University of Wisconsin. All these papers were preliminary, and Kimura did not want them in his collected papers except for the paper on the "stepping stone" model, revising Wright's model of his shifting balance theory, in No. 3 (1952).

In the *Annual Report* No. 2 for the year 1951 (1952), Kimura first encountered the "problem" of this book: how does a researcher who knows cytology and chromosomes use them in population genetics? Kimura introduced the controversies of Fisher and Ford vs. Wright and Dobzhansky over the role of "random genetic drift" and selection in natural populations. Here is how Kimura solved the "problem" in 1951:

> Let x be the relative frequency of a gene (or a chromosome type) A, and let $1-x$ be that of its allele (or a different chromosome type) A' in a panmictic and sufficiently large population.... (Kimura 1952, p. 53).

In this passage, Kimura used a "gene" or a "chromosome," indicating he knew he was reducing a chromosome to one locus. In the next paragraph he described the Fokker-Planck equation, discussed by Sewall Wright in 1945, to measure diffusion in a natural population, which means Kimura was interested in random changes of selection of genes in large populations. Wright introduced the Andrey Kolmogorov paper of 1935 in his 1945 paper for the National Academy of Science, and discussed the Fokker-Planck equation, one forward model of Kolmogorov, in his 1949 paper.

Wright's use of diffusion equations of Kolmogorov and Fokker-Planck and its influence on population genetics

Kolmogorov was an outstanding mathematician in the USSR for many years, and deeply devoted to spreading the use of mathematics in other sciences. He studied in Germany in the late 1920's and published his mathematics thesis on diffusion equations in German in 1931, but soon returned to the USSR. He was a leader in applying diffusion equations in mathematics to other fields. Kolmogorov's thesis does not mention Fisher, Wright, or Haldane, for application of his diffusion equations to heredity. His thesis was submitted in 1929 and he did not have either Fisher's 1930 book or Wright's 1931 or 1932 papers, but he clearly read them before publishing a three page paper in English entitled "Deviations From Hardy's Formula in Partial Isolation," directly addressing Wright's shifting balance theory.

> R. A. Fisher and S. Wright mathematical researches [cites Fisher 1930 and Wright 1931] refer to the evolution of the gene's concentration in a population where free crossing is prevalent. The aim of the present work is to indicate that method which would enable us to obtain analogous results in a population disintegrating into partial feebly connected populations. (Kolmogorov 1935, 129)

Kolmogorov hoped his work on diffusion equations would show both Wright and Fisher a better way mathematically to simplify Wright's shifting balance theory. Kolmogorov sent Wright a copy in 1945. Wright cites it in his long paper published in France (Wright 1939), but perhaps the French publisher added it. Wright replied to Kolmogorov immediately in a paper (Wright 1945).

Kolmogorov was clear about his diffusion model. Both N. P. Dubinin and D. D. Romašov, well known geneticists in the USSR and experts in population genetics, told him that many models were consistent with mathematical modeling populations. Kolmogorov could find only one model for mathematical treatment: "A series of equally interesting schemes of limited crossing do not submit, so far, to any mathematic process of treatment" (Kolmogorov 1935, 129). All Kolmogorov could do was to suggest one better statistical method to Wright for his shifting balance model.

Wright refused to answer most papers sent to him with no prior connection, but this was an important paper on his own topic. He started with a strong support for Kolmogorov:

> Dr. A. Kolmogorov has recently been kind enough to send me a reprint of an important paper on this subject which was published in 1935 but which had not previously come to my attention. While the application is restricted to a particular stationary distribution, the method of approach points to more systematic formulation than before. (Wright 1945, 383)

Wright was clearly impressed by this diffusion equation: "in fact, completely general for the stationary form of distribution may be shown by a slight modification of a method that has been used for derivation of [distribution of genes in a population](Wright 1945, 384)." Wright also showed how this distribution could be applied to a steady flux of genes over time. Wright also showed that the equations of Fisher and himself over the time of their theories had some better solutions, but Kolmogorov had a more exact version, when conditions were met. Any reader of this paper would know that Kolmogorov had provided a real mathematical advance to the theories of Fisher and Wright, but only in the conditions that Kolmogorov specified.

Wright's second paper on diffusion equations came in 1949. He introduced the Kolmogorov diffusion equation in talking about population structure and often summarized his views on evolution and pointed to its most frequently used form:

> I am indebted to Dr. L. J. Savage for calling to my attention that this same formula (Fokker-Planck) has been used in physics in the analysis of Brownian and allied types of random movement. (Wright 1949, 564)

Wright held that Kolmogorov had the general case of the equation but it had some limitations, including the same probability for any allele to move from one sub-population to another separate sub-population. Many missed Wright's paper in 1945 in the Proceedings of the National Academy of Sciences (PNAS). Wright's 1949 paper appeared in a popular book published by Princeton University Press and many, like Kimura, read this paper.

Kimura's background in population genetics and statistical physics

Professor M. Kumazawa, Kimura's biology teacher in high school, taught him to care about statistical physics in population biology. Kimura worked almost every day pursuing chromosome structure in lilies in his laboratory. Kumazawa taught his biology course and Kimura was much impressed by his belief in Mendelism and chromosomes: "the explanation that Mendel's first law could be understood in terms of random assortment of paired chromosomes" (Kimura 1985, 461). Kimura knew about others applying quantitative methods to genetics while in the Eighth National High School in Nagoya. He found in one class on using mathematical methods of theories of probability and statistics in science: "I was impressed by the fact that an apparently chaotic phenomenon has laws." His physics teacher conveyed to Kimura that "one can describe natural phenomena in mathematical terms starting from only a few basic principles" (Kimura 1985, 461; shades of R. A. Fisher in 1922, and many physics teachers).

The National High Schools had, like universities, professors with doctorates, but they had small classes and were interested in all the students individually. Kimura spent only 2.5 years there during the war, but was deeply impressed with his education, which "had a decisive influence on forming my professional career for the rest of my life" (Kimura 1985, 460). As a student in Kyoto University, Kimura was a student of botanical cytology, and soon became "more and more absorbed in mathematical treatment of genetical phenomena rather than in observational and experimentation on chromosomes" (Kimura 1985, 464). By 1949, Kimura was copying the papers of Fisher, Wright, and Haldane from the journals available at labs of Kihara (wheat) and Komai (mammals), a habit he continued after he moved to Mishima and the National Institute of Genetics, where Komai had an appointment.

Wright understood that Kolmogorov used a forward version (the Fokker-Planck equation). Before he met any western population geneticist, Kimura knew how to improve the work of Sewall Wright on "random genetic drift." Kimura tried to explain this to Muller who visited the National Institute of Genetics in 1951, but Muller did not understand him. One American population geneticist, Newton Morton, a student of Crow at the University of Wisconsin, came to visit Japan and met Kimura. He simply could not believe that Kimura was so advanced in his attempts to better Wright's views of "random genetic drift" by using more elegant equations by Kolmogorov.

Fisher and Wright both understood that their equations on population genetics were tied to near genetic equilibrium in the population; otherwise, their equations would not follow a population closely. Kimura knew that too, and he knew that Kolmogorov additions would add to constrictions in the Fisher and Wright genetic equilibrium. All believed their use of statistical physics would help to understand populations over time, and that statistical physics was the key.

Kimura's early papers: physics applied to biology

When Kimura realized that Kolmogorov had written this lovely new and shorter version of diffusion equations, he read a new physics textbook by K. Kunisawa, in Japanese, with this English translation, "Modern Probability Theory" (1951). When I came to Japan in 1988, he gave me copies of the crucial pages 202-203, where Kunisawa showed that Kolmogorov had two versions of his equations: forward, like the Fokker-Planck equation, and a backward version, not contained in the Kolmogorov paper of 1935 sent to Wright. Kimura understood its meaning for understanding the fixation of mutations in populations. No one else had this insight, as far as I can tell, at this time in 1951.

Kimura's experience of not convincing Muller made him realize his English was poor, but he continued to think well of Muller. Crow and Wright reintroduced Kimura to Muller at Cold Spring Harbor in 1955. Muller became an encouraging advisor to Kimura and encouraged him to apply population genetics to molecular evolution. Kimura was deeply convinced that Muller was right, and started soon to apply his work to molecular biology.

In 1951, Kimura kept a notebook, "Statistical Theory of Population Genetics 1951," with dates marked many times. Kimura developed a way to understand the forward equation of Kolmogorov as a graphical equation. Kimura developed a deep relationship with Kolmogorov's diffusion equations that would last all his life. He often thought about these equations and often referred to this graphical version as a way of understanding them.

In the National Institute of Genetics Annual Report for 1951, published in 1952, Kimura published two special papers. In the summer of 1953, he revised these two papers into one on board a ship to the USA to study at Iowa State University. When he got to Iowa State, he received an invitation to go to a meeting in Madison, Wisconsin, where Wright was visiting in preparation to move from Chicago after the next academic year. Wright took one look at the paper, and suggested immediate publication after Kimura's revisions. Why would Kimura's two papers written in 1951 together in a single paper turn Wright into such a strong advocate of Kimura after spending only a few hours with him in Madison? Not only was Wright impressed, Crow was also, and volunteered to submit this paper to *Genetics* on September 7, 1953. What had Kimura done to impress both Wright and Crow? I will try to explain what Kimura did in these two papers (not reprinted in the collected papers of Kimura), and how he put them together in one paper that had an immediate effect upon Wright and Crow.

Kimura used the forward Kolmogorov equation, citing it as the Fokker-Planck equation, to model the issues in random changes in the selective gene frequencies for the locus in question in a very large population. He preceded this paper with one that did the same thing for heterotic genes or cyclic genes. He began with a very large panmictic population in the first article and introduced the paper:

> In the mathematical treatment of the process of change by selection, it has commonly been assumed that the selection intensity is constant throughout many generations, though in practice the intensity must be subject to fluctuations. In the present investigation random fluctuation of the selection intensities is assumed to occur and the resulting Markov process [the Fokker-Planck equation] is investigated. (Kimura 1952, p. 56)

As the process goes on, fixation of alleles at the locus does not disappear in this diffusion process. The process was very different than Wrightian "random genetic drift," where alleles are lost as the population gets smaller. Kimura concluded:

> This process should proceed indefinitely, if there were no disturbance by gene mutation, though in practice the advance will finally be checked by the opposed mutation pressure. (Kimura 1952, 57)

The random changes in selection rates led to change over much time; some alleles were almost lost, and others almost fixed.

In the second paper, *On the Process of Decay of Variability due to Random Extinction of Alleles* (Kimura 1952), Kimura wrote an introduction:

> In a population of restricted size, the relative frequencies of genes undergo random changes from generation to generation owing to the random sampling of gametes in reproduction, so that a gradual

> decrease in genetic variability of the population is to be expected, unless variants are constantly supplied by mutation or migration. (Kimura 1952, 60)

Kimura waited for the time when all hetero-allelic classes were decreased to a state of equilibrium in which they lost variability at the rate of 1/2N per generation. He focused his paper on the stage before equilibrium using Kolmogorov equations, and when the loss reached the stage of loss at the rate of 1/2N, Wright's "random genetic drift" would cause loss of variation.

Narayuki Takahata, hired by Kimura, left these two papers out of the Kimura papers (Kimura 1994). Neither one by itself says much, and Kimura says nothing about the two papers made into one at the later time. That was wholly due to Kimura, who saw that putting these two papers together was impressive to population genetics. Wright had reduced Kolmogorov to the 1/2N model in 1945, and very deftly. Kimura was after more. When he put these two papers together, the result gave population genetics more power for "random genetic drift," and pleased both Wright and Crow.

Kimura's 1954 paper

Kimura believed that his work on "random genetic drift" was different than Wright or Fisher. What Kimura wanted to do on board the ship to the USA was tie changes in selection rates of alleles, creating a continuous surface, to Wrightian "random genetic drift" caused by "random sampling of gametes," and present a "unified way" of looking at "random genetic drift" in natural populations by using two mathematical approaches. His approach certainly appealed to Wright and Crow.

The title of Kimura's 1954 paper was misleading: 'Process leading to quasi-fixation of genes in natural populations due to random fluctuation of selection intensities." I say misleading only because Kimura compares the results of "random fluctuation of selection intensities" to "random sampling of gametes" and shows how both are necessary for showing "random genetic drift" in large size and small size, even of the same population at different in times. He was certainly correct that little regard had been paid to Wright's insistent attention to random fluctuations of selection with regard to a single locus. This is so difficult to measure in a natural population that no one actually did it, whereas "random sampling of gametes" could be done by any researcher in genetics, as we saw in Chapter Four of this book.

Kimura begins the paper by stating:

Among factors that may produce random fluctuations of gene frequencies in natural populations, random sampling of gametes and random fluctuations of selective intensities may be especially important in relation to evolution. (Kimura 1954, 280)

Fisher, Wright, Haldane, Muller, Dobzhansky, and many others had paid close attention to "random sampling of gametes." The Kolmogorov forward equation or Fokker-Planck equation showed that this diffusion equation produced "random genetic drift" at a single locus and piled alleles near the zero or one level, but did not eliminate any alleles:

> ... the random fluctuation of selective intensities by itself cannot lead to the complete fixation or loss, in the strict sense, of the gene contrary to the case of random drift due to small population number.... In other words, after a sufficient number of generations almost all populations will be in such a situation that the gene is either almost fixed in the population or almost lost from it. To distinguish this from fixation or loss in the case of small effective population number, the terms "quasi-fixation" and "quasi-loss" are proposed. (Kimura 1954, 289)

In any longer time, the alleles did not disappear: "In the long run, this process of quasi-fixation or -loss will be checked by the opposing mutation pressure" (Kimura 1954, 289).

Kimura put together his two papers by the following method. He made a diagram of quasi-fixation and quasi-loss of alleles using the forward Kolmogorov diffusion equation, with its continuous surface, a major influence on allele frequencies at the locus under question. What Kimura showed in this article is that using the Kolmogorov forward equation applied to a large population. He created a graph that showed how effective population size could be minimized and he could show where Wright $1/2N$ generations to fixation or loss could be seen on the graph. He was trying to show that the Kolmogorov forward equation could point toward small populations where Wright's "random genetic drift" could be found. Both of these processes can work in a natural population, said Kimura.

Kimura, for the first time in population genetics, could talk about "random genetic drift" in a single population from large size (using "random genetic drift" with Kolmogorov diffusion equations applied to selection fluctuations of allele frequencies) to small size (using "random genetic drift" with 1/2N). In large populations, Wright's "random genetic drift" means little, Kimura said, but random fluctuations of selection in the locus could drive the population to easily produce "quasi-fixation" and "quasi-loss" and clearly influence allele frequencies at the locus in question. Total fixation or loss would be impossible. In smaller populations, random fluctuations of selection at the locus mean little, and Wright "random genetic drift" could produce in 1/2N generations many fixations or loss of alleles at a locus. Kimura had done in this paper what Fisher, Wright, and Haldane had been unable to do: to see these disparate processes both working in every population, depending on effective population size.

Wright and Crow were impressed because both wanted "random genetic drift" in all sizes of a population. Both were looking for ways to increase the importance of "random genetic drift" in evolution. Both were interested in this brave young man who came from Japan to study in the USA. Kimura put Kolmogorov's diffusion equation to great use for a form of "random genetic drift" but not to fixation, using a surface. He also left room where Kolmogorov diffusion could not work well for the usual Wright "random genetic drift" in small populations. Kolmogorov and Wright equations went together, but each had its realm of effect in Kimura's paper.

According to Kimura, small populations had Wright's "random genetic drift," and large populations had Kolmogorov surfaces produced by shifting selection rates of alleles. Population geneticists, led by Wright, did not adequately distinguish inbreeding and "random genetic drift." What Kimura had done in this 1954 paper was to tie together Kolmogorov and Wright methods by using different sizes of population. Soon Kimura would have another way to tie Kolmogorov and Wright views without the problems in this paper.

Kimura's rewritten 1954 paper in 1955

Kimura tried in his above paper to use the Kolmogorov equation to its fullest. He made little progress on it during his year at Iowa State. In the summer of 1954, he moved to the University of Wisconsin and, even before the school year, began to work again on how to relate the forward Kolmogorov equation with Wright's "random genetic drift." Now he knew some better quantitative ways to relate them than in his 1954 paper.

He had two clues. Wright had used the forward Kolmogorov equation in one of his papers with Kerr, though he had just 8 (4 males and 4 females) members in each population. Wright was not confused about showing "random genetic drift" using the forward

Kolmogorov equation in a small population. Wright used it because the Kolmogorov equation worked more easily with his model of "random genetic drift." Kimura discovered that Wright had found a way, in this model, to completely relate Wright's "random genetic drift" and Kolmogorov's graphical approach, but only in the case of Kerr's data. Wright had no way to relate them generally so one could see the quantitative relationship of the forward Kolmogorov equation and Wright's "random genetic drift." Wright's ability to use this relationship with Kerr data stimulated Kimura to work on the general problem. If successful, then Kimura would have solved the problems of his 1954 paper of having two different analyses of "random genetic drift" in differing population sizes.

The second clue came from the Ph.D. thesis of S. Goldberg, who in 1950, in an unpublished thesis at Cornell University, showed that Kolmogorov's forward equation could be used to display the gene frequency of Wright's "random genetic drift" if one also assumed recurrent mutations. "His solutions have a direct connection with the frequency distribution of unfixed classes of pure random genetic drift" (Kimura 1955a, 145), according to Kimura, who demonstrated how to turn Kolmogorov's forward equation into the gene frequencies of Wright's "random genetic drift," or go the opposite direction.

Kimura now knew a better interpretation of Kolmogorov's forward equation that could work across all population sizes. Kimura could easily go from small to large populations or from large populations to small ones. He used either Wright's version of "random genetic drift" or Kolmogorov's surface equations for any population size. Fisher's mathematics fit easily into the Kolmogorov's equations. Kimura united Fisher's and Kolmogorov's equations into a single whole. Wright was enormously pleased and submitted it to the PNAS. Wright had been unable to apply Kolmogorov's forward equation to "random genetic drift" in small populations, and told the audience at the Cold Spring Harbor meeting in 1955 that Kimura had accomplished a task he simply could not solve (see Wright's comments after Kimura's presentation in Kimura 1955b).

Problems came from the different kinds of surfaces for calculating Wright's "random genetic drift" using Kolmogorov equations. The small kinds of differences in Kolmogorov equations, mapping the genes on a continuous surface, was possible for Kimura. Wright's "random genetic drift" focused upon small populations well. A modern population genetics book presents both a 2-dimensional version of Fisher/Wright and a 3-dimensional surface based upon Kolmogorov, or a complete agreement of the Kolmogorov surfaces and gene frequencies versions (Hartl and Clark 2007, pp. 108-109).

Kimura published his thesis (Kimura 1955b), and wider use of Kolmogorov equations (Kimura 1964). He published a paper on selection modeled on Fisher's Fundamental Theorem, where he used Kolmogorov equations. Fisher insisted he had already worked this out, but was happy to publish Kimura's paper. Fisher added to his paper, "I already did this" (Kimura's letters). Kimura worked on island models. And he worked on "random genetic drift" and selection.

Before 1968, Kimura became the most famous mathematical theorist in the world in population genetics. He never escaped the work of Fisher, Haldane, and Wright, and he enlarged their influence in population genetics and later his neutral theory.

The neutral theory of molecular evolution

In 1967, James F. Crow wrote a paper on J.B.S. Haldane's changing belief in how quickly a population responded to selective forces in the wild (Crow 1968). Crow was busy at Wisconsin at this time with a deanship and other duties and so wrote the paper as soon as possible, and also sent copies of this paper to Wright and Kimura. One passage greatly interested Kimura. Crow assumed that only 10,000 genes coded for proteins in most organisms, perhaps 1/10 of the number of genes held in the genome. Kimura wrote out his paper two times before having it typed and then sent copies to Crow and Wright. Kimura wrote to Wright a handwritten letter with his paper with this phrase: "I am very serious on my present note." Kimura wanted to include the entire neutral genome in his neutral theory. Neutral alleles required no costly selective adjustment to evolve: just lots of evolutionary time.

Wright, who professed no interest in the non-coding DNA, answered Kimura in three pages. Wright appeared convinced by Kimura's argument but thought it unimportant for evolution. He thought that Kimura was going to be the master of DNA that did not matter in evolution. Crow did not answer, and explained to me later that he was very busy with an administration job at Madison at this time. After six or seven drafts, Kimura submitted the paper to *Nature*. James Maddox, the editor, with conflicting reviews, accepted Kimura's paper as "Letters to the Editor," published in February, 1968.

An unusual conclusion in Kimura's paper was the disappearance of effective population size from his equations. Kimura was arguing that even in large populations, "random genetic drift" would continue to make neutral evolution work. Kimura believed that small populations still had the most "random genetic drift."

Kimura's paper was explosive in evolutionary biology. Proceeding through evolution, Kimura argued most evolution occurred in the 90% "unused" part of the genome. He also included any coding DNA that could not be distinguished in the genome. Wright cared only about evolution in the part of the genome that made a difference in the population. Kimura had found a way to evolution for the other 9/10's of the genome.

Everyone in evolution knew the DNA might be the native stuff of evolution, but in 1968 they had a problem. Geneticists had begun to sequence protein molecules that were made in sexual populations, but biologists had no sequences of DNA. That would not be part of evolution until 1983-1985, when DNA sequencing began to blossom. Kimura's neutral theory had no major sequences of DNA to use until after his 1968 paper, nor even in the book he wrote on his theory in 1983 (Kimura 1983). This book was in press before many DNA sequences were available. Without the sequences of neutral DNA, the neutral theory was having a tough time by using protein data, produced by maybe 10% of the DNA.

I was working as a graduate student with Dick Lewontin when he and Jack Hubby were working on sequencing proteins in natural populations of *Drosophila pseudoobscura*. He thought his protein research was "right next to DNA." After Kimura published his neutral theory, with protein data but meant for neutral DNA sequences, Kimura had a problem. These sequences of proteins did not conform to his completely neutral model. So he had to change a lot about his neutral theory until the DNA sequences were available.

Part of Kimura's difficulties came from his newly hired colleague, Tomoko Ohta. She joined his work on the 1968 neutral theory, but quickly became involved with applying a theory to proteins instead of unavailable DNA sequences. Having no DNA sequences was tough for Kimura, who tried to make his neutral theory simple–he loved that in his 1968 paper. The beautiful simplicity of his neutral theory appealed to him. Ohta saw his theory, based upon protein data, and modified his model to better fit the protein data. Ohta, to fit her data, used a computer to calculate her work. Ohta images the protein data in her early view (Ohta 1973) as "slightly deleterious" mutations which would, of course, not be Kimura's completely neutral model, but certainly a model that would appeal to many evolutionists who disbelieved Kimura's neutral theory would work with DNA.

A famous paper on the neutral theory known well to Crow, Kimura, and many others, was published in 1969, based entirely upon protein data. This paper was written before Kimura's paper, but not published then. Jack Lester King and Thomas H. Jukes published their paper in 1969, with this title: "Non-Darwinian Evolution: Random Fixation of Selectively Neutral Mutations" (King and Jukes 1969). Kimura disliked the title, but was happy to have this support. King and Jukes, who insisted upon the title, soon would have a problem of extra attention from evolutionists who insisted on evolution by natural selection as the key to evolution in nature.

King and Jukes soon became more famous than Kimura for the neutral theory during the time between their paper and Kimura's book on the neutral theory (Kimura 1983). King was doubtful about Kimura's completely neutral theory, and believed his theory and Ohta's slightly deleterious theory were closely related. Ohta's theory gained a lot of credibility during this time, but she and Kimura also published many papers together on the neutral theory.

One of the most interesting issues was the old theory of attachments to any desirable character that humans selected or nature selected. Darwin was sure of this from his work on domestic or natural animals and plants. Any breeder would cross two breeds and then inbreed to make a new breed, and linked characters would appear. Breeders always got more than they wanted. Wright followed this theme as a breeder of guinea pigs. Linked characters appeared in molecular evolution papers by (Kojima and Shaffer 1967) and by (Smith and Haigh 1974) whose paper bore the title, "The hitch-hiking effect of a favorable gene." Favorable genes always came with other genetic material, according to all four.

Ohta and Kimura answered them in a paper (Ohta and Kimura 1975) that put hitch-hiking of neutral alleles to rest: "The effect of selected linked locus on heterozygosity of neutral genes (the hitch-hiking effect)." Crow read the paper and made comments for presentation.

The final discussion section contained the crucial statements to explain why the protein data seemed less present than predicted from the completely neutral theory. The first argument was that linked characters (what they call linkage disequilibrium) were never a factor in large populations.

> Together with our previous studies on linkage disequilibrium (Ohta and Kimura, 1971b; Ohta, 1973), our analyses indicate that linkage is important only in small and transient populations such as those at the time of speciation, and not in large and stable populations. In small populations, linkage disequilibrium due to random drift may have significant effects on the behavior of surrounding genes. However, the average effect is rather small even if the individual effect gets rather large. In this respect, linkage only makes chance effects somewhat larger in transient populations. Lewontin (1974) seems to be overemphasizing the role of linkage, especially since, except where inversions are observed, linkage disequilibrium is rarely found (Mukai, Watanabe, and Yamaguchi, 1974). For large and

> stable populations, the concept of quasi-linkage equilibrium (Kimura, 1965; Nagylaki, 1974) together with the single locus theory is sufficient to treat most problems realistically. (Ohta and Kimura 1975, p. 325)

I see this paragraph as the work of Kimura. His insight into the problems of molecular evolution told him that the best way to work is by using single locus in his equations and treating 90% of DNA as being neutral. Chromosomes were fragile, constantly changed, and a category not worth working into the equations. Ohta did not agree with this bald policy of using just single locus models. Yet, she did that too in the next paragraph, and stated her theory:

> Maynard Smith and Haigh (1974) have argued that the adaptive gene substitutions at many loci, simultaneously occurring, may reduce the heterozygosity at other loci drastically if the population size is large. They suggested that this may explain why the observed heterozygosity per locus is not much different among various species. Unfortunately for the neutral theory we have to deny their conclusion and state our belief that the total size of the species is the most important parameter that determines the amount of random drift. Then we have to search for another explanation for the uniformity of average of heterozygosity among various species. It is possible, as proposed by one of us (Ohta,

> 1974) that the very slight negative selection based on functional constraints of the protein molecule becomes effective in very large populations leading to mutation-selective balance at many loci and this prevents the level of heterozygosity from increasing indefinitely as the population increases. On the other hand, random drift prevails in relatively small populations. (Ohta and Kimura 1975, 325-6)

Kimura and Ohta agree in this paper that chromosomes can be ignored (although both used them on Mendelian heredity), and one locus models and population size determined "random genetic drift," but they would come to disagree about Kimura's 1968 neutral theory.

Nothing changed about the neutral theory when Kimura's book of 1983 was in print. DNA, however, was beginning to be sequenced. Kimura found in DNA the proof of his 1968 theory, and he moved immediately to DNA sequence data.

Kimura developed his own nearly neutral theory of molecular evolution (Kimura 1979) that applied to protein sequences. Kimura kept the variation as small as possible in the neutral theory by inventing a gamma distribution that kept close to the zero axis, which Kimura compared with Ohta's wider exponential distribution. This same argument was in his 1983 book using protein data.

When Kimura got the DNA sequences in 1986, they fit well to his neutral theory, and he jettisoned his version of the nearly neutral theory. The nearly neutral theory was no longer necessary to his neutral theory, which accounted for almost everything at the

DNA level, as it was known in 1986. He went back to his completely neutral molecular evolution at the DNA level that he believed in since 1968. Kimura loved his 1968 theory of neutral evolution; the outcome with DNA sequences gave him much confidence in his neutral theory of DNA evolution.

Further development of the neutral theory of evolution 1986b

I never saw Kimura happier than after he published the 1986b paper (Kimura 1986b). I came to Japan to meet him again in 1988. He had rediscovered his 1968 theory, and its new setting in DNA gave evidence that fit his neutral theory.

In his last years Kimura had gained the highest honors of an evolutionist. His neutral theory of evolution seemed to have huge support, and he spent most time pointing to how neutral evolution could lead to adaptive evolution, even in the 10% of DNA that seemed to code for something. Kimura was at the top of his profession until his death in 1994.

Kimura believed 90% of the DNA was neutral, but concerning percentages, beliefs with time have changed. Most biologists now believe that 70% of DNA codes for RNA, of so many kinds. For sexually breeding populations, mRNA is edited by an RNA editor that can make many different changes into ten or hundreds of different kinds of proteins. Many of those who study RNA think it existed prior to DNA and possibly evolved into using DNA for evolutionary history (James Darnell 2011). The amount of neutral DNA has now decreased by a huge amount; perhaps 20% of neutral DNA is left, probably less. More than 80% of DNA is occupied by coding at various times in life. Kimura's thesis, neutral DNA, is getting smaller than he thought possible.

Kimura's theory of molecular evolution required huge amounts of time, greater than the time of the species in the fossil record. The neutral theory predicted measuring evolution over speciations rather than thinking about one species. The same is true in Ohta's nearly neutral theory. Testing their neutral theories is tough; extended time is the essence of neutral theories.

Ohta used population size as a crucial variable in all her versions of nearly neutral theory applied to proteins. She believed constantly in population size as the arbiter of "random genetic drift." She supported Fisher, Haldane and Wright that larger populations had little "random genetic drift" and smaller populations had much more. All the problems with Fisher, Haldane, and Wright are present in Ohta's work.

The neutral theory, nearly neutral theory, and "random genetic drift"

The neutral theory and nearly neutral theory missed the supposed biological origin of the "random genetic drift." Neither Kimura nor Ohta saw where "random genetic drift" originated in population genetics. They inherited the theories of Fisher, Haldane, and Wright. Kimura and Ohta often used the phrase "random sampling of gametes" as the source of "random genetic drift." This process produced maximum genetic variation from recombination in meiosis in large populations, and minimum in small populations, and no "random genetic drift." Both thought of evolution over speciation; 10,000,000 generations are not much to either theory.

Having no "random genetic drift" in evolution harms the neutral theories. No matter how we approach the neutral theories, "random genetic drift" is the crucial variable, and does not exist. I can see no way to preserve the neutral theories in population genetics.

The disappearance of chromosomes in the neutral theories is another problem. Population genetics must learn how to deal with chromosomes, which are crucial in evolution. Meiosis is the process that recombines chromosomes and makes gametes. The amount of variation produced by meiosis in a large population is enormous. Kimura and Ohta encouraged population geneticists to dismiss the chromosomes. Population genetics must move towards inclusion of chromosomes.

Chapter Six

Topics in Population Genetics

The argument in the first five chapters can be applied to many areas of population genetics. This book could be many thousands of pages when applied evenly over evolutionary biology, but historians of science and colleagues in evolution will have to deal with all these questions. What I can do in this last chapter of this book is to address some issues not addressed adequately before but fundamental in evolution. The five sections of this chapter are:

1) "Random genetic drift" in prokaryotes
2) "Gene pools," "random genetic drift," and population genetics
3) Selection at a locus
4) Laws of physics, beauty, and "random genetic drift"
5) Fisher, Haldane, and Wright on physics and population genetics

"Random genetic drift" in prokaryotes

Eubacteria and Archaea, the prokaryotes, with only one chromosome, originated 3.5 to 3.7 billion years ago. Eukaryotes, like us humans, left fossils 1.7 billion years ago, but might have evolved at the same time as Eubacteria or Archaea or later before their appearance in the fossil record.

Fisher, Haldane, and Wright used a Mendelian locus. Wright said in the 1930s that no "random genetic drift" occurred in prokaryotes. Population geneticists now share a belief in "random genetic drift" in prokaryotes. Some argue for a Fisher/Wright locus in prokaryotes, but that causes confusion because Eubacteria and Archaea have only one chromosome.

Others evaluate "random genetic drift" in prokaryotes by studying recombination of chromosomes in both Eubacteria and Archaea. These creatures just split to create another, but they do have many ways to recombine with other prokaryotes and eukaryotes. Recombination is a whole science of three mechanisms, each with many possibilities: transformation, conjugation, and transduction.

Transformation requires a bacterium or Archaea to pick up genetic material, test it for the DNA machinery in the cell and many other mechanisms including the RNA, and incorporate the loose genome if it fits.

Conjugation requires Eubacteria or Archaea to transfer to another genetic organism some or all of its genome. Eubacteria can mate with another Eubacteria but the process is very complicated. When Eubacteria couples with an Archaea, which they have done from the beginning, mating is more difficult. Matings between Eubacteria and Archaea still happen now, despite having 2.5 billion years since they parted in evolution from each other. Humans have some Eubacteria genome in our DNA; a single person carries ten times more Eubacteria cells than human cells. The organism getting the DNA must have ways to deal with this genetic gift.

Finally comes transduction, which means the use of bacteriophage (bacterial viruses) to conduct the transfer of genetic material. This genetic variation must fit into the existing DNA and RNA.

In each case geneticists know rather little in evolutionary time. Geneticists follow transfer of genetic material in all three ways; they call it recombination in Eubacteria and Archaea. Geneticists know that Eubacteria and Archaea do not practice meiosis as do eukaryotes.

Recombination used by prokaryotes is useful for geneticists speaking about "random genetic drift." The same thing happens in eukaryotes, but more happens. In meiosis, one duplication of the pair of chromosomes, recombination, and two reductions of the 4-way of chromosomes, yields four gametes. For each pair of chromosomes, every resulting gamete would be different because the reductions do not give the same gametes. Eukaryotes in meiosis practice recombination many more ways than the prokaryotes. Eubacteria and Archaea do recombine and we saw from eukaryotes that no "random genetic drift" occurs in recombination.

Those who study prokaryotes apparently had no trouble in attributing recombination, or many other arguments, as the cause of "random genetic drift". Those who study "random genetic drift" in eukaryotes attribute it to inbreeding, like Fisher and Wright, or to the cumulative accidents of sampling of gametes. None of these causes worked to make "random genetic drift" in eukaryotes. Do the same causes in prokaryotes produce "random genetic drift?"

Prokaryotes do have recombination. The eukaryotes have no "random genetic drift" from recombination. Does recombination that occurs in prokaryotes produce "random genetic drift?" No, recombination does not produce "random genetic drift" in either prokaryotes or eukaryotes.

"Gene pools," "random genetic drift," and population genetics

"Gene pools" were missing from the origin of population genetics. I did not use this term in my book on *The Origins of Population Genetics* (Provine 1971), which ends in early 1932 with the work of Fisher, Haldane, and Wright. Dobzhansky heard his colleagues talk about "gene pools" in Russia, but barely understood them. Dobzhansky later invoked them for his version of population genetics in 1951a and 1951b, as we mentioned earlier in Chapter Three. Published as among the most read publications, Dobzhansky's keynote speech of the 50th anniversary of the discovery of Mendelian heredity was published in *Genetics*, and the other was the third edition of his famous book, *Genetics and the Origin of Species*, the most used book in evolution courses for 20 years. In both works he continually used these words: "*A Mendelian population is, then, a reproductive community of individuals which share in a common gene pool.*"

"Gene pools" can mean two different things. 1) Population geneticists use "gene pools" to focus upon a single locus with many alleles on a chromosome. By showing a jar holding all the alleles (using colored balls) at a locus (usually 2 alleles), geneticists show that random choice of alleles will produce "random genetic drift." They demonstrate "gene pools" and "random genetic drift" with jars in textbooks and encyclopedias all over the world. 2) Population geneticists use "gene pools" to demonstrate genetic variation stored in an entire population, as Dobzhansky mentioned above. Population geneticists use this version of "gene pools" to enable their field to go forward or backward in time, previously impossible.

A population geneticist can use both definitions in a single population, at the level of locus or at the population level. Indeed, population geneticists often show you the locus version, and then say this occurs at every locus in the population, as Wright often did.

Dobzhansky soon lost control the phrase, because population geneticists discovered that having a "gene pool" was a huge gift to them. The "gene pool" was the addition to population genetics that solved a lot of technical problems. The use of this concept grew beyond anything imagined by Dobzhansky. Population geneticists do not mention Dobzhansky when they use "gene pools." Any introductory biology text presents "gene pools," especially when they talk about elementary population genetics.

Wright did not use "gene pools" in his writing until they were put into population genetics by Dobzhansky and Crow in the 1950s, who pointed to Fisher and Wright as the fathers of using "gene pools." A population of the right size to all loci varying is Wright's ideal for changing the genetics of the population over time. After the 1950s, population geneticists used the term "gene pools" profusely, and now use the "gene pool" in populations to talk about "random genetic drift."

Crow used these one locus models for his entire working life. He has equally defended "beanbag" genetics, "gene pools," "random genetic drift," and natural selection, as enormous allies to population genetics and evolution. In 2001, Crow published in *Nature* a thoughtful defense of these concepts in his article "The beanbag lives on" (Crow 2001, 771). He stated: "The gene-pool model is a wonderful, simplifying convention" that leads directly to Mendelian heredity and the Hardy-Weinberg law. He adds:

> Random sampling from a small parental population leads to random gene-frequency drift, because in the draw some alleles are over-represented, others under-represented, or not at all. These models lead to the simple mathematical equations of which classical population genetics largely consists. (Crow 2001, 771)

Crow explains later in this paper, "gene-pool models are indeed simplified" and leads to "meaningful simplicity" that sweeps away "disorderly complexity" (Crow 2001, 771). Crow saw the "gene pool" as the best way to teach population genetics.

In 2000, the University of Chicago Press asked me to write an Afterword for the new edition of my book on the history of population genetics. Since writing the first edition (Provine 1971), I had become skeptical about one-locus models, "random genetic drift," "gene pools," and simple Mendelian heredity for talking about alleles at a locus and for a population. I argued in the Afterword that "the notion of 'gene pools' now strikes me as one of the most artificial concepts of population genetics" (Provine 2001, 201).

The simplification of inventing "gene pools" is easily explained. Turning a population into a "gene pool" enables a researcher to use the entire population in population genetics. By using "gene pools" as a codeword for all the loci in the population, any population geneticist generalizes from a one-locus model to all loci in the entire genome of the population. Chromosomes disappear in this process, but this is strange, because the chromosomes come back as Mendelian inheritance in the one-locus model! Using chromosomes for Mendelian inheritance but throwing chromosomes away when generalizing one-locus models to "gene pools" in a population is a self-contradictory way to proceed in population genetics.

My objection to one-locus models in a "gene pool" comes from the assumption that one locus can be extended to all loci. One locus cannot be extended to all loci in the population by assuming "gene pools." Any sexual organism inherits the recombined chromosomes in meiosis of its mother and father, not from the "gene pool." Fisher invented "random genetic drift" by extending population genetics to statistical physics. Wright did not understand how poorly Fisher's thesis applied to inbreeding. Population genetics would improve dramatically by not using "gene pools" anymore.

The appearance of "gene pools" in the work of Russian population genetics in the 1930s built upon the models of Fisher, Haldane, and Wright, who did not use that terminology. Crow, for example, argued that "gene pools" were central to the Fisher 1922 paper, Haldane's papers and book of 1932, and Wright's 1931 paper. Their attachment to the "gene pool" was part of their use of the laws in statistical physics. No wonder Crow would love "bean bag" genetics. When Crow invoked the "gene pool," he had instantly given up on chromosomes for statistical physics. Chromosomes need to be part of population genetics to add great meiosis variation in chromosomes to the population.

Teaching population genetics is tied now to "gene pools," both applied to a chromosome with alleles or to an entire population. Both methods of teaching "gene pools" can be found in books on evolution and encyclopedias, or on the internet. "Gene pools" are inadequate when applied to the chromosomes because meiosis just recombines the chromosomes from both parents, not a "gene pool." Applied to populations, the "gene pools" became an advantage in population genetics, but "gene pools" do not exist in biological populations.

Selection at a locus

Geneticists often talk about selection at a locus. Invention of a locus has many issues, given our discussion of a locus that demonstrates "random genetic drift." Whole chromosomes are crucial in evolution; geneticists know that fundamentally, but population genetics ignores them.

A single locus problem on selection is just like a neutral locus. Fisher put that locus F on the chromosome and concentrated his attention to selection at this locus. The next generation, this locus might easily change from recombination before meiosis. This was unknown to Fisher. Geneticists did not know the incredible function of meiosis in producing variation in the population. The chromosomes are complete in any organism and they change every generation, but organisms keep the chromosomes in the same biological order. The chromosomes can change order over evolutionary time, of course, but no "random genetic drift" occurs.

Meiotic recombination is the source of huge variability in large populations, and smaller variability in small populations. When the population gets small, most chromosomes in the population have disappeared. The population has no "random genetic drift," but does have inbreeding that probably will end the life of the population.

Accepting the chromosomes into population genetics would be a major advance. In sexual organisms, a fantastic system of recombination and sorting chromosomes comes every generation. Population geneticists can model that system, though we are nowhere close yet. As we understand more, we can make population genetics better than we have now, and take meiosis and the entire chromosome into account.

Understanding chromosome mechanics to trace a locus of DNA that recombines every generation is a complicated process. Population geneticists have to understand the whole chromosome, which recombines every generation, and to model it with statistical physics as much as possible. Fisher's model F of 1922 should not be the foundation of population genetics. Population geneticists must invent a new science of their theories of evolution.

Chapter Six

Laws of physics, beauty, and "random genetic drift"

Population genetics became applied statistical physics of evolutionary biology and domestic populations. The article "The contribution of statistical physics to evolutionary biology"(Vladar and Barton, 2011) gives the authors' view in *Trends in Ecology and Evolution*, a premier site for many readers about quantitative evolution. Vladar is a young biologist well trained in physics and mathematics, and Barton one of the best known population geneticists in the world. They begin the paper with this introduction:

> Evolutionary biology shares many concepts with statistical physics: both deal with populations, whether of molecules or organisms, and both seek to simplify evolution in very many dimensions. Often, methodologies have undergone parallel and independent development, as with stochastic methods in population genetics. Here, we discuss aspects of population genetics that have embraced methods from physics: non-equilibrium statistical mechanics, traveling waves and Monte-Carlo methods, among others, have been used to study polygenic evolution, rates of adaptation and range expansions. These applications indicate that evolutionary biology can further benefit from interactions with other areas of statistical physics; for example, by following the

> distribution of paths taken by a population through time. (Vladar and Barton 2011, 424)

How has statistical physics benefitted the science of population genetics? A first assumption is that heredity is basically simple and thus statistical physics can be applied to populations of organisms:

> A similar averaging over equivalent microstates is made in both population and quantitative genetics: one averages over individual gene combinations to describe a population by its allele frequencies, and can further average over all the allele frequencies that are consistent with a given mean and variance of a quantitative trait. In this sense, physicists and evolutionary biologists both model populations (a gas or a gene pool) rather than precise types(individual particles or genotypes). This 'statistical' description in terms of a few variables, the macrostates, summarizes the many possible configurations of the microstates (degrees of freedom), which cannot be accurately measured or described. Furthermore, the macrostates are then sufficient to predict other properties without reference to the microstates. For example, thermodynamics describes macroscopic properties without referring to individual particles; similarly, quantitative genetics does not refer to allele frequencies to predict the trait mean in the next generation. (Vladar and Barton 2011, 424)

"Gene pool" is the key to this assumption. A population of organisms, whether sexual or asexual, was essentially a "gene pool." From the "gene pool," you can push or pull whatever you wish to make statistical physics apply to populations.

R.A. Fisher applied statistical physics in 1922 to inbreeding, and turned "random genetic drift" at a neutral locus of one chromosome into a measure of inbreeding. Fisher took a chromosomal problem (inbreeding) and turned it into a genome problem (a neutral locus). Fisher's thesis became an important part of population genetics. Vladar and Barton describe Fisher's use of "laws of physics" or "gas laws" to turn population genetics into a similar science. They especially like the later Kolmogorov version of diffusion equations, as did Wright and Kimura. All was built upon the model of Fisher, and Wright, who was trying to duplicate the inbreeding used by the Hagedoorns. Fisher's attempt in 1922 was a huge success in population genetics, except for more mathematical development, and further limitations like the help of Kolmogorov's equations. Fisher's thesis has never changed since 1922, yet his attempt to quantify inbreeding was biologically wrong: "random genetic drift" at one genetic locus (genic) did not reproduce inbreeding (chromosomal) in wild rats, especially in small populations. Fisher was unable to specify the biological source of "random genetic drift."

The diffusion equation (for genetic variation), the coalescent process (for history of genetics), and the path ensemble (Wright's approach) all describe the same process and are, under the same conditions, mathematically equivalent. Each has different advantages and limitations. Whereas the diffusion equation, coalescent process, and path ensemble do not require detailed balance in work of a population geneticist, the stationary distribution of all three does require balance. This solution is exact, quite general and relatively simple, and ignores chromosomes.

Crow understood this perfectly (Crow 2001). He described how the "gene pool" has continued its life in population genetics, and continues the science of population genetics. When we turn heredity back into history (the coalescent process), we get the same thing: backward on the "gene pool." The path ensemble of Wright generates the same result. They are all "the same process and are mathematically equivalent" according to Vladar and Barton (Vladar and Barton 2011, 425) for all three things: futures of populations, history of populations, and path ensemble of populations. What an achievement in biology achieved by statistical physics, but the limitations are huge.

The amazing thing about the application of statistical physics is that Fisher apparently "got it right" in 1922. Otherwise, population geneticists would have challenged him again and again. I strongly disagree with Fisher's understanding of "random genetic drift," yet never has Fisher's application of statistical physics to population genetics changed. Population genetics now is the same as in 1922. Fisher's invention of F is false, and Wright and Haldane and so many others have made this same mistake in population genetics.

Physicists use their statistics to make predictions about the future. We can do this with gases or any statistical process in physics or chemistry. We can start with the inverse relationship between two bodies and make good predictions about where they will be in the future. The statistical process, not precisely the truth about the process, applies well in a statistical sense to many problems. Can we use population genetics to predict the future of a biological population the way physicists and chemists do in inanimate molecules? No one even pretends to be capable of doing this with populations beyond saying that small populations easily become extinct.

Population geneticists can only predict the past, and a vast amount of genetic information disappears. Population geneticists should be skeptical about combining statistical physics and population genetics. Vladar and Barton thought of this view, as did Fisher and Wright before them:

> Statistical physics is based on universal physical laws. By contrast, biological concepts are relative, plastic, or even arbitrary (e.g. mean fitness or traits). Hence, the analogies with statistical mechanical models are limited, depending on the nature of epistasis, physical linkage of the genes, unpredictable fluctuating selection, and so on. Moreover, there are different ways in which precise analogies can be drawn, limiting their scope: some factors act deterministically (e.g. selection) and others stochastically (mutations or drift). (Vladar and Barton 2011, 430)

Notice that the authors disagree completely with Fisher's thesis that natural selection was indeterministic. Population genetics pays little attention to epistasis, and RNA biology is very active in many ways including epistasis. More of the DNA on a chromosome is devoted to RNA than anything else. Chromosomes mean little to population genetics because they are complicated and change over time and hard to put into statistical physics. Vladar and Barton set for themselves this question in the last paragraph: "Of course, we can ask whether the mathematical paraphernalia that we advocate is of any practical use" (Vladar and Barton 2011, 430). The rest of the paper is just a summary of the paper, suggesting that applying statistical physics to population genetics makes a lot of sense. Their paper praises population genetics. They do accurately specify precisely why the models of population genetics often do not behave as the same models applied to physics.

The problem with applying statistical physics to natural populations comes by addressing chromosomes. Understanding what happens to recombined chromosomes is the key to biology today. Why does population genetics wish to disregard chromosomes? Because statistical physics maps evolution easier without chromosomes. Thus Fisher was the center of population genetics in 1922, and retained that position until the present. Population genetics now is Fisher's production.

I followed "random genetic drift" in the work of Fisher, Haldane, and Wright when writing my thesis on the origins of population genetics, and later I wrote a long book on Wright

(Provine 1971, 1986a). After interviewing Sewall Wright I did not even begin to doubt "random genetic drift," nor would any student in classes of population genetics or evolutionary biology because Fisher, Haldane, and Wright all believed in "random genetic drift."

My argument challenges "random genetic drift," but also raises questions about the many problems of population genetics without revision. Fisher made a biological mistake, but he had no idea how much he would lead population genetics astray. He supposed that statistical physics would be a fine quantitative basis for evolution of populations, or population genetics, by inventing the F on a chromosome. Fisher chose indeterminism because he wanted to have free will for humans and the beautiful simplicity of using few variables and statistical physics for population genetics. What I object to in Fisher is not just his conclusion of "random genetic drift," but the use of statistical laws of physics to model populations subject to inbreeding, a condition that results in loss of chromosomes.

Fisher, Haldane, and Wright on physics and population genetics

When Fisher, Haldane, and Wright were young population geneticists in the 1920s, all three were in their 30s, and loved the physics that was being made at the same time. They embraced the complete integration of physics, chemistry, and biology into one science. Fisher, Haldane, and Wright did not invent this idea. Physicists and chemists did this job for them. Some of the physicists were religious, and then gods or other intelligent agents were responsible for life. Fisher's deep religious views, Haldane's dialectic materialism with a model of mind with free will, and Wright's belief in monistic panpsychism and free will claim few advocates in population genetics now.

In 1989 the outstanding population geneticist, Marcus W. Feldman, edited a book entitled *Mathematical Evolutionary Theory* (Feldman 1989), with distinguished professors of population genetics. The index listed two non-informative entries on "chromosome." The same word was used other times and not listed in the index. Mathematical population genetics appeared to have nothing in common with actual chromosomes. Mendelian inheritance is there but not there. Every participant would deny slighting chromosomes even though they have disappeared in population genetics. Population genetics must not ignore chromosomes.

Sewall Wright had the toughest time with "random genetic drift." He imaged the "kaleidoscopic" changing at every locus from "random genetic drift" in small populations, or a large population divided. His career would have been helped by giving up this concept. He could have kept his "shifting balance theory of evolution" in natural populations with just inbreeding, as he did for farmers who read his manual (Wright 1920). He would have been left with inbreeding only in smaller populations, and this would have been to his advantage in many ways. He was always tied up with having inbreeding and "random genetic drift" together in small populations, same as Fisher and Haldane. Wright nevertheless would have to give up Fisher's attempt to place statistical physics into biology, and Wright disliked this idea.

For large populations, divided into smaller populations, Wright thought every locus would begin "random genetic drifting." All Wright needed to explain his view was an understanding of meiosis, which provided all the genetic variation he could possibly want, but which he never saw. Meiosis has no "random genetic drift." Meiosis produces recombined chromosomes that become gametes. Gametes produce genetic variation but no "random genetic drift" (Chapter Two).

Cytologists began to interpret meiosis by 1940, but their work did not remake population genetics. Wright was wedded to his policy of connecting inbreeding and "random genetic drift." Fisher, Haldane, and Wright maintained to their deaths no change in their views, and modern population geneticists have agreed with all three.

Biology now is fundamental in the sciences. Biophysics and biochemistry are great sciences that largely have replaced both physics and chemistry in biology. What has happened to population genetics? Like physics and chemistry, population genetics also has had a loss of faculty in recent times. Back in the 1920s, physicists had a great science and everyone wanted their sciences to look like physics. Population genetics was invented by Fisher in 1922, and never altered by population geneticists to reflect increased understanding of actual causes of evolution, especially meiosis. Now we know a lot about meiosis. Population genetics is unreal in modern biology.

References

Adams, M. B. 1979. From "gene fund" to "gene pool": On the evolution of Evolutionary Language. *Studies in history of biology* 3:241-285.

Baker, W. K, Gregg, T. G., Neel, J. V. and Stalker, H. D. 1975. Warren Spencer (1898-1969) *Genetics* 79:1-6.

Birdsell, J. B. 1950. Some implications of genetical concept of race in terms of spacial analysis. *Cold Spring Harbor Symposia on Quantitative Biology.* 15:259-314.

Boveri, Th. 1903. Über die Konstitution der chromatischen Kernsubstanz. Verh. D. Zool.Ges. 13. Würzburg, Germany.

Boveri, Th. 1904. Ergebnisse über die Konstitution der chromatischen Substanz des Zellkerns. Jena. G. Fischer.

Box, J. F. 1978. *R. A. Fisher: The life of a scientist.* New York: John Wiley & Sons.

Buri, P. 1956. Gene frequency in small populations of mutant Drosophila. *Evolution* 10:367-402.

Crow, J. F. 2001. The beanbag lives on. *Nature* 409:771.

Crow, J. F. 1968. The cost of evolution and genetic loads. *Haldane and Modern Biology*. Ed. K. R. Dronamraju. Baltimore: The Johns Hopkins Press.

Crow, J. F. and Morton, N. 1955. Measurement of gene frequency drift in small populations. *Evolution* 9, 202-214.

Darnell, J. 2011. RNA: Life's indispensable molecule. Cold Spring Harbor, NY. *Cold Spring Harbor Laboratory Press*.

Dobzhansky, T. 1937. *Genetics and the origin of species*. New York: Columbia University Press. 2nd ed., 1941. 3rd ed., 1951a.

Dobzhansky, T. 1951b. Mendelian populations and their evolution. In *Genetics in the 20th century*, ed. by L. C. Dunn, 573-590. New York: Macmillan Company.

Dobzhansky, T. 1970. *Genetics of the Evolutionary Process*. New York: Columbia University Press.

Dobzhansky, T. and O. Pavovsky. 1957. An experiment of interaction between genetic drift and natural selection. *Evolution* 11:311-319.

East, E.M. and D.F. Jones. 1919. *Inbreeding and outbreeding: their genetic and sociological significance.* Philadelphia: J.B. Lippincott.

Feldman, M. W., ed. 1989. *Mathematical Evolutionary Biology.* Princeton: Princeton University Press.

Fisher, R. A. 1918. The correlation between relatives on the supposition of Mendelian inheritance. *Transactions of the Royal Society of Edinburgh* 42:399-433.

Fisher, R. A. 1922. On the dominance ratio. *Proceedings of the Royal Society of Edinburgh* 42:321-41.

Fisher, R. A. 1924. The biometrical study of heredity. *Eugenics Review* 16:189-210.

Fisher, R. A. 1930. *The genetical theory of Natural Selection.* Oxford: Oxford University Press.

Fisher, R. A. 1934. Indeterminism and natural selection. *Philosophy of Science* 1:99-117.

Fisher, R. A. 1950. The creative aspects of natural law. *The Eddington memorial lecture.* Cambridge: Cambridge University Press.

Glass, H. B., Sacks, M. S., Jahn, E. F., and Hess, C. 1952. Genetic drift in a religious isolate: An analysis of the causes of variation in blood group and other gene frequencies in a small population. *The American Naturalist* 86:145-159.

Glass, H. B. 1953. The genetics of the Dunkers: In the hands, ears, and blood of the members of this small religious sect a geneticist finds evidence for an important force of human evolution: random drift. *The Scientific American* 189:76-81.

Glass, H. B. 1954. Genetics changes in human populations, especially those due to gene flow and genetic drift. *Advances in genetics.* 6:95-139.

Gulick, J. T. 1888. Divergent evolution though cumulative segregation. *Journal of Iinnaean Society of London* (Zoology) 20:189-274,

Hagedoorn, A. L. and A. C. Hagedoorn. 1921. *On the relative value of the processes causing evolution.* The Hague: Martinus Nijhoff.

Haldane, J. B. S. 1932. *The causes of evolution.* London: Longmans, Green and Co.

Haldane, J. B. S. 1934. Quantum mechanics as a basis for philosophy. *Philosophy of Science* 1:78-98.

Haldane, J. B. S. 1957. The cost of natural selection. *Journal of Genetics* 55:511-524.

Haldane, J. B. S. 1960. More precise expressions for the cost of natural selection. *Journal of Genetics* 57:351-360.

Hartl, D. L and Clark, G. L. 2007. *Principles of Population Genetics*, 4th edition. Sunderland, MA: Sinauer Associates, Inc. Publishers.

Hedrick, P. W. 2000. *The Genetics of Populations*, 2nd edition. Sudbury, MA: Jones and Bartlett Publishers.

Huxley, J. S. 1942. *Evolution: The modern synthesis*. Oxford: Oxford University Press.

Huxley, J. S., ed. 1940. *The New Systematics*. Oxford: Oxford University Press.

Jones, D. F. 1917. Dominance of linked factors as a means of accounting for heterosis. *Genetics* 2:466-79.

Jones, D. F. 1944. Biographical memoir of Edward Murray East, 1879-1938. *National Academy of Sciences, Biographical Memoirs* 23:215-242.

Kerr, W. E., and Wright, S. 1954a. Experimental studies of the distribution of gene frequencies in very small populations of *Drosophila melanogaster*: I. Forked. *Evolution* 8:172-177.

Kerr, W. E., and Wright, S. 1954b. Experimental studies of the distribution of gene frequencies in very small populations of *Drosophila melanogaster*: III. Aristapedia and spineless. *Evolution* 8:293-302.

Kimura, M. 1951. 5 reprints in one folder. *Annual report of national institute of genetics.* 1:45-51.

Kimura, M. 1952. 7 reprints in one folder. *Annual report of national institute of genetics.* 2: 53-66.

Kimura, M. 1953. 3 reprints in one folder. *Annual report of national institute of genetics.* 3:62-66.

Kimura, M. 1954. Process leading to quasi-fixation of genes in natural populations due to random fluctuation of selective intensities. *Genetics* 39:280-295.

Kimura, M. 1955. Solution of a process of random genetic drift with a continuous model. *Proceedings of National Academy of Sciences*. 41:144-150.

Kimura, M. 1964. *Diffusion models in population genetics*. Methuen's review series in applied probability, volume 2.

Kimura, M. 1968. Evolution rate at the molecular level. *Nature* 217:216-218.

Kimura, M. 1979. A model of effectively neutral mutations in which selective constraint is incorporated. *PNAS USA* 76:3440-3444.

Kimura, M. 1983. *The neutral theory of molecular evolution*. Cambridge: Cambridge University Press.

Kimura, M. 1985. Genes, populations, and molecules: a memoir. In *Population Genetics and Molecular Evolution*, ed. by Ohta, T. and Aoki, K., 459-481. Tokyo: Japan Scientific and Societies Press.

Kimura, M. 1986a. Diffusion Models of population genetics in the age of molecular biology. In *The craft of probabilistic modeling: A collection of Personal accounts*, ed. Gani, J., 151-165. New York: Springer-Verlag.

Kimura, M. 1986b. DNA and the neutral theory. *Philosophical Transactions of Royal Society of London.* Series B. 213:343-354.

King, J. L. and Jukes, T. H. 1969. Non-darwinian evolution. *Science* 164:788-798.

Kojima, K. and Shaffer, H. E. 1967. Survival process of linked mutant genes. *Evolution* 21: 518-531.

Kolmogorov, A. 1935. Deviations from Hardy's formula in partial isolation. Comptes Rendus (Doklady) de l'Académie des Sciences de l'URSS. 3 (8) No. 3(63):129-132. Sewall Wright has this reprint very marked with his comments. In possession of the Rare Book section of the Cornell Library.

Lack, D. L. 1945. *The Galapagos Finches (Geospizinae): a study in variation.* Occasional Papers, no. 21. San Francisco: California Academy of Sciences.

Lack, D. L. 1947. *Darwin's Finches.* Cambridge: Cambridge University Press. Reprinted in 1983.

Lack, D. L. 1973. David L. Lack: Obituary. *Ibis* 115:421-441.

Ludwin, I. 1951. Natural selection in *Drosophila Melanogaster* under laboratory conditions. *Evolution* 5:231-242.

Mayr, E. 1942. *Systematics and the Origin of Species.* New York: Columbia University Press.

Mayr, E. 1954. Change of genetic environment and evolution. In *Evolution as a process*, ed. J. Huxley, A. C. Hardy, and E. B. Ford, 157-180. London: Allen and Unwin.

Merrell, D. J. 1953. Gene frequency changes in small laboratory population of *Drosophila melanogaster*. *Evolution* 7:95-101.

Ohta, T. 1973. Slightly deleterious mutant substitutions in evolution. *Nature* 246:77-78.

Ohta, T. and Kimura, M. 1975. The effect of selected linked locus on heterozygosity of neutral alleles (the hitch-hiking effect). *Genetical Research* 25: 313-326.

Planck, M. 1932. *Where is science going?* With a preface by Albert Einstein. New York: Norton Press.

Prout, T. 1954. Genetic drift in irradiated experimental populations of *Drosophila melanogaster*. *Genetics* 39:529-545.

Provine, W. B. 1971. *The origins of theoretical population genetics*. Chicago: University of Chicago Press. 2nd edition, 1991.

Provine, W. B. 1985. The R. A. Fisher-Sewall Wright controversy and its influence upon modern evolutionary biology. Oxford Surveys in Evolutionary Biology 2:197-219.

Provine, W. B. 1986a. *Sewall Wright and evolutionary biology*. Chicago: University of Chicago Press.

Provine, W. B. ed. 1986b. Sewall Wright Evolution: selected papers. Chicago: University of Chicago Press.

Provine, W. B. 1988. Progress in evolution and meaning in life. In *Evolutionary progress*, ed. M. H. Nitecki, 49-74. Chicago: University of Chicago.

Provine, W. B. 2005. Ernst Mayr, a retrospective. *Trends in Ecology and Evolution* 20:411-413.

Provine, W. B. 2006. Ernst Mayr: Genetics and speciation. *Genetics* 167: 1041-1046.

Reed, S. C. and Reed, E. W. Natural selection in laboratory populations of Drosophila. *Evolution* 2:176-186.

Shull, George Harrison. 1908. The composition of a field of maize. *American Breeders' Association* 4:296-301.

Smith, J. M., and Haigh, J. 1974. The hitch-hiking effect of a favorable gene. *Genetical Research* 23: 23-35.

Spencer, W. P. 1935. The non-random nature of visible mutations in Drosophila. *The American Naturalist* 59:223-238.

Spencer, W. P. 1940. On the biology of Drosophila immigrans Sturtevant with special reference to the genetic structure of populations. *Ohio Journal of Science* 40, no. 6:345-361.

Spencer, W. P. 1947. Genetic drift in a population of *Drosophila immigrans*. *Evolution* 1:103-110.

Stebbins, G. L. *Variation and Evolution in Plants*. New York: Columbia University Press.

Stern, C. 1950. *The principles of human genetics*. San Francisco: W. H. Freeman and Co.

Stimuzo, M. 1984. My life with lilies. *The lily yearbook of the North American lily society*. 37:16-27.

Sutton, W. S. 1902. On the morphology of the chromosome group in Brachystola magna. *Biological Bulletin*: 4:24-39.

Sutton, W. S. 1903 The chromosomes in heredity. *Biological Bulletin*: 4:231-251.

Takahata, N. ed. 1994. *Population Genetics, molecular evolution, and the neutral theory. Selected papers of Motoo Kimura.* With a forward from James F. Crow. Chicago: The University of Chicago Press.

Vladar, H.P., and Barton, N.H. 2011. The contribution of statistical physics to evolutionary biology. *Trends in Ecology and Evolution* 26:424-432.

Wagner, M. 1868. Die Darwin'sche Theorie und das Migrationsgesetz der Organismen. Leipzig: Duncker and Humblot.

Weinstein, A. 1980. Cytology in the T. H. Morgan School, In *The Evolution Synthesis*, ed. E. Mayr and W. B. Provine, 80-86. Cambridge: Harvard University Press.

Wood, A. J., Severson, A. F., and Meyer, B. J. 2010. Condensin, and cohesin complexity: the expanding repertoire of functions. *Nature Reviews/Genetics* 11:391-403.

Wright, Sewall. 1920. *Principles of livestock breeding.* United States Department of Agriculture Bull. 905. Washington, D.C.

Wright, Sewall. 1922a. *The effects of inbreeding and crossbreeding on guinea pigs*. I. Decline in vigor. II. Differentiation among inbred families. USDA Bull. 1090.

Wright, Sewall. 1922b. *The effects of inbreeding and crossbreeding on guinea pigs*. III. Crosses between highly inbred families. USDA Bull. 1121.

Wright, Sewall. 1929a. Evolution of dominance: comment on Dr. Fisher's reply. *American Naturalist* 63:556-61.

Wright, Sewall. 1929b. Evolution in Mendelian Populations. *Anatomical Record* 44:287.

Wright, Sewall. 1930. Review of *The genetical theory of natural selection* by R. A. Fisher. *Journal of Heredity* 21:349-356.

Wright, Sewall. 1931a. Statistical theory of evolution. *Journal of Statistical Association* 26:201-208.

Wright, Sewall. 1931b. Evolution in Mendelian populations. *Genetics* 16:97-159.

Wright, Sewall. 1932. Roles of mutation, inbreeding, crossbreeding, and selection in evolution. *Proceedings of the sixth international congress of Genetics* I:356-366.

Wright, Sewall. 1939. *Statistical genetics in relation to evolution.* *Actualités scientifique et industrielles* 802. Exposès de Biometire et de la statistique biologique, XIII. Paris: Hermann & Cie.

Wright, Sewall. 1940a. Breeding structure of populations in relation to speciation. *American Naturalist* 74:232-248.

Wright, Sewall. 1940b. The statistical consequences of Mendelian heredity in relation to speciation. In *The New Systematics*, ed. Julian S. Huxley, 161-183. Oxford: Oxford University Press.

Wright, Sewall. 1945. The differential equation for the distribution of gene frequencies. *Proceedings of the National Academy of Sciences* 31:383-389.

Wright, Sewall. 1949. Adaptation and Selection. Ed. by Gepson, G. L., Mayr, E., Simpson, G. G. *Genetics, Paleontology, and Evolution,* 365-389.

Wright, Sewall. 1964. Biology and the philosophy of life. *The Hartshorne Festschrift, Process and Divinity.* Ed. by W.R. Reese and Eugene Freeman. Chicago: Open Court.

Wright, Sewall. 1968. *Evolution and the genetics of populations, vol.1: genetics and biometric foundations.* Chicago: University of Chicago Press.

Wright, Sewall. 1969. *Evolution and the genetics of populations, vol. 2: the theory of gene frequencies.* Chicago, University of Chicago Press.

Wright, Sewall. 1977. *Evolution and the genetics of populations, vol. 3: Experimental results and evolutionary deductions.* Chicago: University of Chicago Press.

Wright, Sewall. 1978a. *Evolution and the genetics of populations, vol. 4:Variability within and among natural populations.* Chicago: University of Chicago Press.

Wright, Sewall. 1978b. The relation of livestock breeding to theories of evolution. *Journal of Animal Science* 46:1192-1200.

Wright, Sewall. 1988. Surfaces of selective value revisited. *American Naturalist* 131: 115-123.

Wright, Sewall and Kerr, W. R. 1954. Experimental studies on the distribution of gene frequencies in very small populations of *Drosophila melanogaster.* II. Bar. *Evolution* 8:225-240.

Made in the USA
San Bernardino, CA
31 January 2015